Into Africa
With a Smile

Linda Bootherstone

Copyright © 2015 Linda Bootherstone

All rights reserved.

ISBN- 978-1517109905
ISBN- 1517109906

CONTENTS

Dedication

Acknowledgements

Maps

Poem   Africa

Preface

Chapter 1	Making Decisions	1
Chapter 2	Preparations	5
Chapter 3	Morocco	9
Chapter 4	Into the Desert	18
Chapter 5	Algeria	22
Chapter 6	Tamanrasset	26
Chapter 7	Agadez	30
Chapter 8	On Through the Desert	34
Chapter 9	Kano, Nigeria	40
Chapter 10	On to Chad	45
Chapter 11	Central Africa Republic	49
Chapter 12	Bangui	52
Chapter 13	Into the Congo	58
Chapter 14	Kisangani	63
Chapter 15	Through Zaire to Rwanda	69
Chapter 16	A Long Way Round, Wisely	74
Chapter 17	Rwanda	78
Chapter 18	Tanzania	82
Chapter 19	Nairobi	88
Chapter 20	Mt Kenya	91
Chapter 21	The Kenyan Coast	98
Chapter 22	Ancient Cities	102
Chapter 23	An Enforced Stay	106
Chapter 24	Into Zambia	110
Chapter 25	Hippo Mine	115
Chapter 26	The Road to Victoria Falls	118
Chapter 27	Rhodesia	124

Chapter 28	Rhodesia's Foundation	127
Chapter 29	Salisbury	130
Chapter 30	To South Africa	139
Chapter 31	The Lay of the Land	142
Chapter 32	Downhill to Durban	145
Chapter 33	Durban Delights	148
Chapter 34	Life in Westville	151
Chapter 35	More Durban Delights	155
Chapter 36	A Make-over for the Bike	159
Chapter 37	The Coastal Route	165
Chapter 38	Cape Town	168
Chapter 39	Father of the Nation	171
Chapter 40	Matters of the Heart	174
Epilogue		182
Preparation		183
Poem	An African Odyssey	191
The Author		
Other Publications		

DEDICATION

For the many travellers with whom I have crossed paths over the years.

ACKNOWLEDGMENTS

When writing about events that occurred many years in the past I could not rely just on my own memories, so needed to check with other people who were in Africa at that time. In this endeavour I have been fortunate enough to reconnect with a few old friends who have been kind enough to help me with information and photographs.

I give my heartfelt thanks to the following people:
John Morgan (diaries and Central Africa photographs)
Jane Thomas (nee Mathews)
Devon Dold
Gene Visser
Jacqueline Griffin
Bill Snelling (Isle of Man records)

Photos in North Africa, Kenya and South Africa are mostly my own but my films of Central Africa were lost so many photographs of this area are courtesy of John Morgan. In some cases I have had to use photos from the public domain to illustrate other places in the narrative.

Book Cover Design: Linda Bick
Rear Cover Photo: Sally Frost
Editing: Viola Weidmann
Publishing Assistance: Mary Gudzenovs

Route Through Africa

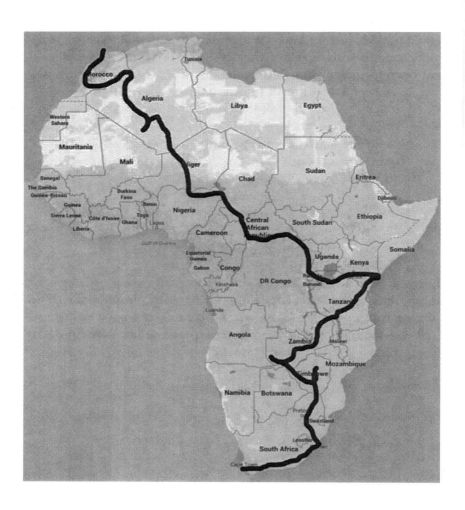

Africa

You have shown me your beauty and your turmoil.

I have been shocked by your violence and soothed by your tenderness

I have revelled in the magnificence of mountains and wide vistas of valleys, lakes, and rivers

In dense forest I glimpsed the fleeting brilliance of butterflies and heard screaming monkeys and roaring lions

I have wondered at rain-bowed waterfalls crashing into deep chasms, sunsets blazing over blood-red earth or sinking slowly into western bays

Your teeming towns and cities are alive with many-coloured people, languages, and songs

Your heart is a beating drum sending its message to the world and bewitching me

PREFACE

The air was hot and sticky, only to be expected in this tropical climate. Thick vegetation each side of the track prevented any view of the surrounding countryside; it was like riding through a never-ending tunnel of green. Hour upon hour I had been battling the rocky, potholed surface that passed for a road. For a little while I managed to pick up speed to 25 mph and then came to a halt surveying a particularly rough patch through which to pick my path.

A rustle came from the bushes and then, suddenly, a dark figure rushed out into the open, running towards me. The man was semi-naked with flashing eyes and a mop of wiry African hair. He approached waving a panga and screaming. I sat frozen, astride my heavily-laden motorcycle. On this particularly rough stretch I could not just open the throttle and go. Heart beating wildly I sat, awaiting my fate as the man came closer, gurgling unintelligible words.

As calmly as I could I fixed a broad smile on my face and thrust forward a gloved hand in a gesture of friendship, 'How do you do? My name is Linda. I'm so pleased to meet you and it's a lovely day isn't it?'

The African stopped his gesticulating with the lethal looking implement and ceased his cries. He looked at me with an expression of pure amazement then turned and disappeared into the jungle.

Shaking like a leaf I fixed my attention back onto the track ahead and gingerly began a tentative forward motion.

'What on earth am I doing here?' I asked myself.

CHAPTER ONE

Making Decisions

Yes indeed, what was I doing here in the middle of Africa alone on my 18- year- old R50 BMW 500cc motorcycle?

Africa. The word had always conjured up ideas of adventure and mystery – steaming jungles, pygmies with their bows and arrows, huge, wasted desert lands with Arabs on their camels, many maybe wrong impressions given by movies such as Africa Queen with visions of retired colonels staring rheumy- eyed into their whisky or gin and tonics muttering, "Ah yes, when I was in the Congo…"

Since 1963, when I first discovered the joys of motorcycling, my itchy feet and throttle hand had taken me all over the British Isles, a fair amount of Europe and Scandinavia, and a short way into the USSR for an FIM (Federation of International Motorcyclists) rally in Moscow. After that there was a three-year tour around Australia and it was in that vast continent that I first experienced long distances in arid lands and learnt to ride on unmade roads.

I returned to England from Australia in 1972 as my father was ill and my siblings had written to say that it could be serious. For a while he seemed to be on the mend and I began to look around for another adventure. Africa appealed. I made enquiries at the AA (Automobile Association) about a journey through that continent and their advice when I mentioned it would be by motorcycle was "Don't!" That was somewhat discouraging so I turned my attention to America. Plans for this were well underway when my father took a turn for the worse and in a short time died, so all ideas of travelling were postponed while my brother and sisters and I helped our mother face her loss. By the time we had her settled, the summer months were over and it was too late in the season to contemplate travelling in the northern hemisphere so my thoughts once again turned toward Africa.

Whilst living at home in the UK I was busy holding down three jobs to save money for whatever journey I would take next. During the day I worked as a driver for a spare parts firm in Thornton Heath – Tridon Spares owned by Don Dew. He was a lively character, expert on all British Leyland cars and a true motor enthusiast. Two or three nights a week I worked in a local pub, The Royal Oak, on the London to Brighton Road, and over the weekend I slaved over an old Singer treadle sewing machine that my father had taught me to use. On this I made suede waistcoats (lined), which I sold, mainly to my folkie friends.

My one regular social event was performing at the South Croydon Folk Club held in the Swan and Sugarloaf pub on Friday nights. As a singer, accompanied by my lagerphone (a percussion instrument I had learnt to play in Australian bush bands) I was soon accepted and dragged onto the committee. We frequently booked guest artists and invited such luminary names as Ewan McColl and Peggy Seeger, Frankie Armstrong and Martin Wyndham Reed. The craic was good and I even learnt to play a guitar during this period.

An interesting incident that occurred at this time involved one of my suede waistcoats. Although I usually made these items for sale, I indulged in making a blue and light grey one for myself and sewed a cloth BMW badge on the front. It looked very smart. It so happened that BMW advertised an open day at their showrooms in Park Lane, London. They had released a new 900cc model and were offering test rides. A special guest was Mike Hailwood, the famous motorcycle World Champion. As I had myself raced while in Australia and followed the sport both there and in the UK, I was keen to meet him and also ride one of the new model bikes. So I arrived at the very flashy showrooms, where a great deal of well-dressed, important-looking people were circulating among the tea and biccies, and took my turn for a ride on the new model. It was impressive and a mark-up from the 600cc R60/5 I had owned in Australia (but I still preferred the old basic Earls forks models pre 1970s.)

Looking smart in my new waistcoat with its BMW emblem, I thoroughly enjoyed a chat with Mike Hailwood. In his early thirties he was of medium build with a trim, fit body, high forehead with a slightly prominent nose and a very open, friendly smile. Mike was a legend in his own time. Fondly called 'Mike the Bike' he had an impressive record. Fortunately born into a wealthy family, who already owned a motorcycle dealership, he was not lacking in good machinery to race but he soon proved to be an exceptional rider.

His first race was in 1957 at Oulton Park and his performances then and later were such that by 1961 he was racing for Honda and in that year was the first man in history to win three races in one week at the Isle of Man TT, the 125, 250 and 500. In 1962 he signed with MV Augusta and won four consecutive World Championships. His greatest rival was the Italian rider, Giacomo Agostini, and they had a memorable battle in the 1967 Senior TT. Mike relocated to South Africa in 1968 and pulled out of GP racing but continued riding at selected meetings in Europe and the UK and entered into car racing, competing in a Le Mans race. In 1973 while competing in the South African GP he collided with Clay Regazzoni and both cars caught fire. After his driving suit had been extinguished by the marshalls, Mike rushed back to help rescue Regazzoni from his still burning car. He later received the George Cross for gallantry.

During my conversation with Mike I told him of my intention to ride down through Africa and he wished me well and said to call in and see his family in Durban if they were there at the time as they had houses in different countries.

After my chat with him I continued circulating and it appeared that my smart waistcoat had given the impression that I was one of the BMW promotion team as I was chatted up by a nice young American man called John. He worked in IT in London, owned a Honda 4 and was looking for information on the new BMW. I couldn't help him with the technical details but was quite happy to accept his invitation to dinner.

But what about my own transport?

I had left my new BMW R60/5 in Australia and only had enough money to buy an old BSA A7. This leaked so much oil that the kick-start was usually too slippery to use so I mainly bump-started it, a skill I had learnt on the race-track in West Australia. As my savings increased I was on the lookout for another old model BMW and found a 1957 R50 going cheap. This became my new love. She had the original separate saddle seats, a tiny back light and was lacking both a side or main stand so, until I had them fitted, I had to lean her up against a lamp-post or shop window.

Now that I had a suitable steed, I needed to find information on how to make this ride through Africa. There was no such thing as the

Internet therefore no overland information sites such as 'Horizons Unlimited', not even Lonely Planet Guides and the AA could only recommend Michelin maps so I began to make enquiries elsewhere.

Chapter Two

Preparations

BMW were considered the best bikes for world travel, a fact of which I was well aware. They were of sturdy German construction, known for their reliability and were used by police and for missionary work in many countries. I had already done many miles on them myself. However, I had no information from any individual who had taken one all the way through Africa so I thought I would pay a visit to the BMW owner's club in London to find out if anyone there had first-hand knowledge of such a journey.

Riding my recently acquired tatty R50 I found the club's meeting place in an upstairs room in North London, lent my steed up against a shop window and glanced at the gleaming new models parked neatly at the curb. I entered the premises wearing an old leather jacket I had found at a car boot sale and repaired on the Singer. It looked a little out of place among the new BMW clothing that most of those gathered were wearing. However the group politely greeted me and listened to my enquiries about information on a trip through Africa. None of the members had made such a journey and they appeared a little sceptical when I told them of my intention. They were helpful enough to promise that they would put my enquiry into their magazine that was circulated to members worldwide. There were a few living in Rhodesia and South Africa and this fact did subsequently prove very helpful.

Apart from this contact with the BMW owners club, I was a member of the Women's International Motorcycle Association (WIMA) and they also had members in Southern Africa whose addresses I was given. But I still wasn't sure what paperwork and preparation I would need. As the AA had been so unhelpful I looked around for another source of information on the logistics.

At this time there were two companies of whom I had heard that arranged overland trips : 'Encounter Overland' and 'Penn Overland' who filled their trucks with approximately 20 passengers and took them on a three-month journey from London to Johannesburg. They were well-known commercial businesses but in a London newspaper I found another private person advertising such an enterprise. I called the number given and spoke to a man by the name of Terry Wilkinson. I explained what I was intending to do and said I needed help with planning.

He kindly invited me to meet him at his home in Barnet, North London. I found the semi-detached house in a quiet street where Terry lived with his mother and she brought in some tea while we had our chat. Terry was a tall, spindly, round-shouldered person in his mid-thirties, I think, with a shock of straggly, dark, receding hair and large round glasses which gave him an owl-like expression. He looked more like an absent-minded professor than an explorer. His usual profession was a part-time insurance contractor but he found arranging overland journeys more stimulating and lucrative.

First he explained how he ran his business. He would buy an old ex-army Bedford truck (3 ton, 4WD) at a vehicle auction and then employ a mechanic to set it up for the journey. This was usually a young Australian used to bush mechanics, on a world tour. While the truck was being prepared Terry advertised for the passengers and got them organised. I explained that I wanted to go by motorcycle but had no idea of the logistics, right from getting visas to what route I should take and what to carry. Terry was keen to help and said that there really was no problem. He had made two successful journeys previously and was just gearing up for another one in February 1974, about six months hence. If I liked, I could come with the truck, at least through the Sahara where they could carry spare fuel and water for me. He also offered to obtain my visas while he was chasing them for the passengers. This would be a great help otherwise I would have to take time off work to go to the many embassies in London and they were notorious for their long queues at the visa departments.

The other document I would need was a 'carnet de passage'. This is a customs document rather like a passport for the vehicle. To save having to pay import duty and then reclaim it every time you entered or left a country, a bond was left with the Automobile Association at the home country and they issued a double-paged document that had to be stamped in and out at the customs posts. This was to ensure that the vehicle did not illegally stay in the country. At the final destination the import duty must be paid on the vehicle or, if returning to its home country, the bond would be returned if all the pages had been correctly stamped. The deposit is usually the worth of the vehicle so it is much better if you have a low value vehicle, such as mine. In the UK at that time I was able to leave an insurance policy as collateral for this. This carnet procedure still applies in many countries around the world and the format depends on the country in which it is issued. Carnets are usually valid for a year and can be renewed in the country they are issued but this is not the case everywhere.

Terry was so helpful and enthusiastic about the forthcoming trip that I told my friend, Jacky Griffin, who had been with me in Australia and she said she was interested in being one of Terry's passengers in February.

Having sorted the paperwork problem I now had to turn to preparing the bike.

I was a member of a local motorcycle club, the Saltbox MCC, located in Biggin Hill, Kent. One of its members, Ronnie Barker, was a BMW enthusiast. He and his wife, Rene, and his cousin, George, with his wife, Doreen, used to come on our club runs with their BMW outfits. Ron was a tradesman and he lived just a few streets away. He kindly agreed to help me prepare my machine for its journey. We wheeled the bike up the path alongside his small terraced house in South Croydon, over the dog-shit strewn lawn (they had two Alsatians) and into his small shed at the end of the garden. This was next to the railway line from Purley to Croydon and we could hear the trains rumbling by.

In this little shed the work began. First we checked the valves and rings and put in a new clutch plate. We also fitted new wheel, fork and steering bearings to try and eliminate the famous BM wobble. I had bought a second-hand dual seat and found a later model tail light. Brackets were made up to fit leather saddle bags and we put on trials handlebars. I thought it would also be a good idea to drill and wire the sump, gearbox and final drive drain nuts to stop them working loose on the bad roads. The piece de resistance was the 6 gallon aluminium tank that my good friend, Geoff Branston, made using a fibreglass one as a model. While living in Australia Geoff, a skilled coach-builder, had made tanks for the racing fraternity in Adelaide.

Geoff lived about 80 miles away and could not see the bike for a precise fitting so when it arrived a week before departure I had to bend the bolts to fit but Geoff had made a fantastic job of it and my boss, Don Dew, had paid for the work as long as I put 'Tridon Spares' in lettering on the side. My very own sponsor!!!

I had a German Harro tank bag, the lower part of which was used as a tool bag and place for spare clutch and brake cables. I thought it wise to carry a spare clutch plate as I thought I might burn one out in the sand. That went in and, because of my valve- dropping experience in Australia, I took spare exhaust valves.

My sleeping bag lived in the top half of the tank bag, something soft to lean on, and the top box on the rear rack was my kitchen with an Optimus petrol stove and pans and space for tins of sardines, pasta etc. The panniers held my clothes, a tow rope and a jerry can. The tent and lilo (inflatable mattress) sat on top of the top-box.

Not long before I left, Ted Simon (later to write *Jupiter's Travels*) had an entire centre spread in the Sunday Times of all the items he would be taking on his now famous journey around the world on a Triumph. It transpired that he left around the same time as me but took a different route through Africa so our paths didn't cross. Ted's journey was covered by his newspaper with reports he sent while I had sporadic articles about my journey in *The Motorcycle* as one of our club members, Gerry Clayton, was writing for it and I sent him information when I could.

So finally, in February 1974, with departure photos taken, somewhat worried farewells from my mother and much excitement from me, I headed south to Southampton to meet the other passengers on Terry's Blue Truck (including Jacky) to take the P &O ferry to Tangiers.

"I had a feeling you were going somewhere," my mother said as she waved me goodbye.

INTO AFRICA – WITH A SMILE

Morocco to Chad

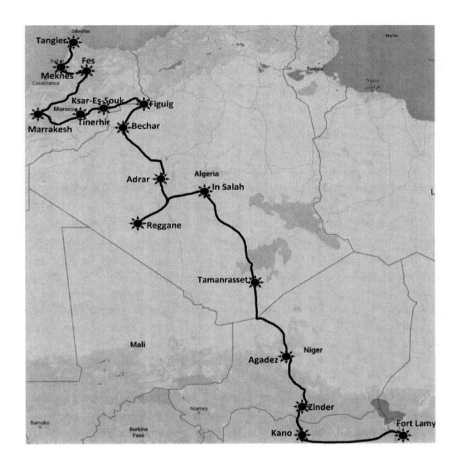

CHAPTER THREE

<u>Morocco</u>

When I left school, in the early 1960s, Britain and most of the Western world was enjoying an economic boom. Jobs were easy to find for both skilled and unskilled workers. However, in the early seventies things started tightening up as we were then hit with an oil crisis. The Arab states put up the price of crude oil and we could no longer afford to use power so lavishly in factories or homes. At one stage the English government imposed a four- day week to save fuel and therefore production was down. Exports were cut and companies suffered. Some bright spark started an 'I'm backing Britain' campaign where employees were encouraged to work an extra ½ hour a day for no pay to keep their employers going. Were we gullible or what! Yes I did. On top of that, the price of fuel went up at the petrol pumps so everyone was complaining.

In February 1974 when we were due to leave the UK and drive down through Europe taking the normal route to Africa via the ferry from Algeciras in Spain, Terry decided that this would be too risky for two reasons. The main one was that he may not be able to obtain fuel on the way through Europe to even get to the first step of the epic African journey. Whereas I had a 300 mile range on my six gallon tank, the truck needed 250 gallons at each stop and if it was unavailable or even rationed, there could be serious problems. That was a risk he was not prepared to take.

The other reason was that it was a particularly bad winter and there was a good deal of snow throughout Europe, especially over the Pyrenees. I was quite happy to eliminate this part of the trip; riding in snow at high altitude was not my idea of fun. So, we were booked on a P+O Ferry from Southampton to Tangiers via Lisbon in Portugal. It would be an amazing transition from our known world in Britain to the Dark Continent in just a couple of days.

Although the ship was well appointed with plenty to see and do, I

Into Africa – With a Smile

am not a good sea traveller and crossing the Bay of Biscay sent me scuttling to lie down in my cabin where I stayed for most of the two-day voyage, worrying about how well my bike had been tied down in the hold. When we docked and unloaded in Tangiers, I found that it had indeed fallen over, had a dented tank, bent centre-stand and smashed ignition switch. Luckily it started and I was able to straighten things out once we had settled into our first campsite outside the city where I got to meet 'the gang'.

The nineteen passengers on the bus were mainly young and single although there were three married couples, one from the UK, one from Australia and one from Canada. They had all paid £200 for the journey that covered their transport and camping food which was bought locally en route. The group formed teams of three for buying local food and cooking it and were rostered to do this. Depending on how good they were at selecting food, which was often different from the known English fare, and how competent their cooking skills, it eventuated that the meals could be classed as tasty, interesting or inedible. They were provided with army tents in which two or three slept on camp beds but some, including Terry and Dennis, the Aussie mechanic, slept in the truck.

The vehicle had a cab in front and the back was open with a tubular frame overhead to take a tarpaulin with side flaps that could be raised or lowered depending on the weather. Seats were longways on the side and there was room to store the passengers' rucksacks beneath them. It was equipped with side racks to carry the camping gear and spare fuel cans.

Most of these young people, mainly from middle-class families with no previous travel experience, had met before at meetings with Terry but it would be interesting to see how they got along with each other in the small confines of the Bedford truck and in their camping arrangements at night. I, of course, was independent of this set up and just observed the friendships, or otherwise, that were forming. The only passenger I knew well was my friend Jacky, a small, slight, 24- year- old with long, fair hair. Jacky, her sister and I had toured together for 2 ½ years in Australia. It was good to have someone familiar to relate to at least at the beginning of this daunting experience.

The first few days were spent exploring the Moroccan cities: Meknes, Fes and Marrakesh.

Meknes, the old capital, had stately palaces and, while we wandered in the old part of the city, we were fortunate enough to meet a friendly, helpful older Moroccan man who showed us around. He wore a white skull cap and we discovered that this denoted that he had been to Mecca and had studied the Koran, thus was well respected among his peers. We were having to speak French most of the time but a surprising number of young Moroccans and children could speak English and other languages, mainly to beg from the tourists or to sell trinkets.

Fes, now a World Heritage city, was fascinating with its hundreds of narrow cobbled streets just wide enough to let a laden donkey pass. It was divided into areas of differing trade items such as silverware, pottery, leather, carpets, clothes, cooking utensils or food.

The rich aroma of spices helped overcome some of the more unsavoury smells that accompany lack of proper sanitation. One of the more overwhelming odours was that of the leather tanneries. Here the hides were treated by being immersed in urine in large ceramic tubs and then agitated by the workers' feet. As we stood on the viewing platforms above, watching this process, the smell drifted up and just about knocked us off our feet. However the sight of the following process when the hides were dipped in brightly coloured dyes of yellow, purple, red and others, was a lot more salubrious.

The high- walled, narrow streets were so confusing and numerous that it was easy to become lost. We found it imperative to have a guide not only to guard against losing our way but also to deter the numerous other would- be guides and touts that pestered any foreigner. In the streets and market squares the traders shouted out to attract our custom. They displayed their remarkable wares, all very exotic and attractive to us. Jewellery, carpets and clothing were offered and the expected bargaining was accompanied by a serving of mint tea. This tea drinking is a ritual; the tea is poured at height from a silver teapot to allow the liquid to bubble into the small glasses. It is polite, and indeed expected, to drink three glasses full, each one sweeter than the other as more sugar is added to the pot. The three servings are to represent one for you, one for the company and one for Allah. It certainly is an acquired taste especially if you are used to not having any sweetener in your normal brew.

The Moroccan traders were dark and attractive with wide grins and flashing eyes. As it was winter many of the men were wearing woollen or mohair jelabas, a long-sleeved and bodied garment with a hood. These are especially suitable for the shepherds in the high mountains where it gets very damp and cold. The female equivalent of this garment, in lighter material, is the caftan. Many Moroccan women use it as a coat when leaving the house. They cover a multitude of sins. Some of these caftans are a dress in themselves, shaped and embroidered. I bought one of white cotton with a blue embroidered design. How I managed to squeeze it into my already packed panniers I'll never know but I used it as a dress for special outings. I was proud of myself as I fell into the swing of unaccustomed bargaining and managed to buy it for about £2.

We travelled on, further south, crossing the spectacular snowfields of the low Atlas to Marrakesh. Here we explored the famous red-walled city and were enthralled by the medicine men, water sellers, jugglers and snake charmers in Jem al Afra square where the night-time stalls with their gas lamps added a certain magic to the performances. The ambience was somewhat spoilt by the coachloads of American tourists with their loud voices and flashing cameras. It was interesting that not only were the rich American tourists there but the young, hippy ones as well. Crosby, Still and Nash had made Morocco popular with their hit song in 1969, *The Marrakesh Express* and it was the 'in' place to go. Many British and American pop stars were exploring the Moroccan coast and even further south in Ghana (*Temma Harbour* sung by Mary Hopkins). Jimi Hendrix's song *Castles made of Sand* may have been inspired by a rock formation off the beach at Essouira.

Although about 90% Muslim, the people of Morocco are moderates and Western tourists were tolerated as long as the women were not too skimpily dressed.

Leaving the city of Marrakesh on the plains, the previous snowfield experience in the Ante- Atlas paled into insignificance compared to the ride I faced going up the windy roads over the High Atlas, the last range of mountains before the plateau preceding the desert. The roads were icy and therefore slippery, with snow piled high on either side and in some places there was a sheer drop to the river valley far below. Even clothed in my leather touring suit and Barbour clothes I was freezing cold and my numb fingers could hardly feel the controls. I had a few nasty moments keeping the bike upright and was relieved when we finally descended out of the ice and snow and came into the dry, barren area which was the forerunner of the dreaded Sahara.

Motor Cycling UK, January 1974

Into Africa – With a Smile

Linda at home

Mike Hailwood

Mike Hailwood racing

Jacky on BM in Tangier campsite with truck in background

Northern Morocco

Linda Bootherstone

Moroccan policeman

Moroccan street scene

Moroccan street

Moroccan market

Water seller, Morocco

Linda's Moroccan caftan

Into Africa – With a Smile

Martin in Morocco

Jacky in Fes

Jacky in Moroccan shop

Marrakesh

Circuit de Lacs, Morocco

High Atlas (above and right)

In the passes, Morocco

Meeting the camels

Festival, Tinehir

CHAPTER FOUR

Into the Desert

Amazing how, on this southern side of the mountains, the scenery had changed so dramatically. The road on which we now travelled was fast and straight with a good surface. The overall impression was of drab, brown earth and mud-brick houses broken by the occasional splash of green where palm trees either followed a river bed (usually dry) or were clumped around an oasis. Black-clad women were sometimes glimpsed near the houses chasing children or goats.

We approached the little town of Tinehir to find garlands and bunting strewn across the road and loud, tinny music blaring from loudspeakers in the street. The truck could not get through the decorations and was obliged to make a detour around the town during which manoeuvre the petrol started evaporating and they were unable to continue. We spent the night camped in a hotel garden. It also appeared that three of the passengers were sick with some type of stomach bug.

Before leaving UK we had all undergone the painful necessity of several injections. At that time a certificate for yellow fever and smallpox was a must at most borders plus it was advisable to be inoculated against typhoid, cholera and hepatitis A and B. It was also a good idea to take anti-malaria tablets for about a month before leaving and during the trip although, as there are a few different types of the disease, these may not always be effective. However, no matter how one tried it was almost inevitable to get stomach trouble which could be as serious as dysentery and the only thing to try and control it, especially while in perpetual motion was Lomatil to bung you up. I had had my inoculations and had also stored anti-biotics and other forms of medicine in my first aid kit hoping that I would not become too ill along the way.

Fortunately by the following afternoon the afflicted had recovered enough to continue and we all started out for Ksar-Es Souk where

Terry was hoping to get to a bank to change more money for petrol. However, there was a three-day bank holiday when we arrived so we decided to camp outside this town for the night and then continue in the hope that the petrol would last until the next.

It poured with rain overnight and everyone packed up miserably the next morning. I started out ahead and hit the first stretch of unmade road. Up until then, all through Morocco, the roads had been sealed and in relatively good condition. Now in a desert region there was much sand that had blown across the road and, with the heavy rain, it had become very soggy. I rode into a pile and became bogged immediately. I was unable to drag the bike out by myself and had to wait for the truck to catch up and someone to help me. The road was very slippery and I was only travelling at about 15mph and battling to keep the bike upright. I took my tank bag off and put it on the truck to see if I had more control without it. Big mistake!

At lunchtime the others suggested that I start off ahead, a few minutes before them and they would catch me up as my slow speed was delaying their progress. So this I did. It had stopped raining, the road began to dry out and I went faster and faster, pleased with my progress but when I looked back there was no sign of the truck. However, I continued and had just negotiated a particularly deep stream when I thought I had better stop and wait for them. After about half an hour with still no sign of them I decided to turn back. Unfortunately my return passage through the stream was not so successful. The bike spluttered to a halt on exit and I knew I needed to take out and dry the plugs. But woe is me! I had left my tank bag with all my tools in the truck and I was helpless without it.

I was terrified. It was getting dark and it was the first time I had been alone in this strange and barren land. There was a village nearby and I didn't want my presence to be known. The people in the area hadn't looked too friendly and I had already discovered that the kids had a nasty habit of throwing stones. Overcoming my panic, I decided to put up the tent to provide some shelter. I had just erected it and was about to crawl inside when I heard footsteps on the road. Heart beating madly I stopped and listened and then heard the familiar cough of my friend Jacky. What a relief!

Joyfully I called out in the darkness as she approached and realised that she was accompanied by the young Aussie couple, Chris and Sue, from the truck. Apparently the truck had run out of petrol about five

INTO AFRICA – WITH A SMILE

miles behind and, upon Jacky's insistence, they had formed this little search party to find me. We spent a very uncomfortable night with four of us crammed into my two-person tent but at least we were altogether and I was so grateful.

Bright and early the next morning Chris and Sue set off back to the truck to collect my tools while Jacky and I stayed to guard the bike. While we were there, sheltering inside the tent, some kids from the nearby village came over and started to ask questions to which we answered as best we could but then to our amazement they started abusing us using filthy language which, even in French, I could understand. When we tried to ignore them they picked up big stones from the side of the road and threw them at the tent in which we sat, huddled together. A concerned Jacky suggested that I put on my helmet to maybe minimise any injury. It seemed stupid to be scared of 10-12 year-old boys but, believe me, they were very aggressive.

During a slight lull between flying missiles we quickly packed up the tent and started pushing the bike, fully laden along the dusty track. Eventually two boys from the truck appeared with my much-wanted tools and within 10 minutes I had drained out the water, dried off the plugs and had the BM fired up and running back to the truck, ferrying my helpers. However, the truck itself was stranded without fuel and it took nearly two days to acquire some as there was only one regular vehicle a day that passed along this stretch of road. Consequently, hitching 60 miles to the nearest town with a jerry can, then back the next day, kept everyone waiting and did little to inspire confidence in the leaders! Eventually enough fuel was obtained to get the truck back onto the bitumen road and on to the Algerian border at Figuig.

Amongst the truck passengers there was a very attractive girl by the name of Houaine Burgess. Despite the necessary water restrictions and the fact that most days were spent getting smothered in dust from the road, Houaine always managed to look fresh and made up. She acted as a magnet to the Moroccan men. Thus, while we were camped at the border, the Chief of Police invited a few of us, including Houaine, of course, back to his place and made us a Moroccan meal he called kief which was meat cooked in spices and served with couscous, a semolina type of grain which is steamed and has a light fluffy texture. I'm not sure what the meat was, it could have been lamb or goat or even camel. The meal was eaten with our hands and a fingerbowl was provided.

Afterwards we were served with so many glasses of mint tea it was just about coming out of our ears and then, much to our amazement, he passed around a joint. It seems that in Morocco the police raid the tourist campsites where they find the hippies, confiscate their stash and then sell it on to other tourists (though not while in their police uniforms) or smoke it themselves. Not being a user myself and wary of any likelihood of misdemeanour, I politely declined.

It was hardly surprising that the police needed to supplement their income. Morocco was a very poor country with an unstable political system. Education was not compulsory and the vast majority of the people were in the low income bracket. It did have a growing tourist industry including winter sports in the High Atlas and there was mineral exploration. The phosphate mining in the south provided its main exports. The fertile coastal strip produced vegetables which were cheap and abundant in the markets to feed the population but most people scraped a living any way they could.

CHAPTER FIVE

<u>Algeria</u>

From the Algerian border we travelled to Bechar, the first main town on our route. We found this place to be uninteresting with not much to offer at all. The goods in the market were badly made and expensive. After the beautiful workmanship of the Moroccan traders and the reasonable prices after bargaining, we found the Algerian products uninviting.

But one thing we found in Bechar that we had missed in Morocco was the communal bath, the baine oriental, that Jacky, Houaine and I went to investigate. As a public bath for the whole town, the hours for ladies were during the day and for men after 5pm. We entered through a small door into a room made of stone and the heating seemed to be under the floor. Around the walls were little basins with a hot tap over each and a gutter running below. The basin had to be emptied with a tin can provided. You lathered up your body with soap (that you had brought with you) then sluiced yourself down from the water in the basin. Not having had a decent wash for at least a week, we revelled in the chance to wash our hair and have a good scrub down. It was hot and steamy and there were about 20 women and children there who took delight in pointing and laughing at us. We were the only white bodies and therefore caused a great deal of interest and amusement.

Bechar marked the start of the Sahara region. Although the road was still sealed, there were sand dunes on both sides and a great deal of sand blowing across the road. The villages around were very primitive; with mud huts and old-fashioned wells, they looked like something out of Arabian nights and , where there were a few palm trees, painted a typical oasis scene. As we were now well and truly away from the cold northern regions the temperature had risen to the low twenties and there was no sign of rain.

I started having problems with a sticking throttle. The sand was getting into the carburettors and jamming the slides. To overcome this I wrapped insulating tape all around the carburettors.

The truck had its first puncture about seventy miles short of Ardrar and we camped in the sand hills that night. Thinking it would be romantic to 'sleep under the stars' which were magnificent in the clear desert sky, I didn't put up my tent but just lay on my lilo and sleeping bag in the open. However, the temperature dropped dramatically and the cool breeze playing on my face kept me awake so finally I erected the tent and crawled inside. I felt much safer and more comfortable inside my little shelter.

We reached Ardrar the next morning and tried to get a permit to cover the next stretch of road to Reggane which was a short cut but the police informed us that due to high winds the road was exceptionally bad with sand and that even the army trucks were getting bogged. It was advisable to back-track about 80miles and cut across the bitumen to In Salah. This we did.

In Salah was a small town which marked the start of the real desert and the unmade road which continued to the other side of the Sahara. We spent a night there and we saw our first scorpion in the sand. Definitely no more sleeping in the open for me! This town was another place with mud-brick, flat-roofed buildings and seemed to be expanding as there was a brickworks there where I noted hundreds of handmade bricks laid out in the sun to dry. We prepared for the next stretch by fuelling up for 300miles and the truck carried two jerry cans for me.

I took a deep breath. This was it. It would be my real test of riding in the sand. The army were building a new bitumen road but we were not allowed to drive on it and had to follow the unsurfaced track alongside. With the full tank the bike was very heavy and I found riding extremely difficult. I became bogged several times and had to wait for the people on the truck to pick me up and push me out. By letting a good deal of air out of my tyres I eventually got the hang of things and the road sporadically improved. I even managed to get up to 30mph at one stage and took my eye off the road for a second to check my handbag was still in place when -whoops- the next thing I was sprawled on the road, the front wheel having hit a bad corrugation. Once more I had to wait for the others to arrive to pick me up but fortunately no damage was done and I continued.

The road was very hard going and it was somewhat disconcerting to see the number of wrecked vehicles along the way including 2CVs and combi-vans. Now instead of sand there were corrugations and they weren't the sort that smoothed out with higher speed. I was being bumped up and down very badly and finding it increasingly difficult to keep the bike on the track. On stopping for a rest, I found the rear suspension units had broken their seals and were therefore giving no damping effect. This was bad enough but a few miles later the front suspension units followed suit and I had no suspension at all! This meant that the frame and my body were taking all the jolts from the road. The mudguards and pannier fittings began to break and I felt as if every bone in my body was breaking too. I carried on for about 50 miles like this, having to stop frequently to get my vision back. I was being jolted up and down so badly that the road ahead was just a blur and I thought my eyeballs would pop out.

Eventually we came to an unexpected petrol pump at the side of the road and a few trucks were stopped there to fuel up. By this time it was getting dark and I was ready to collapse but I realised that I could not ride the bike any further and would have to take this opportunity to arrange alternative transport. The drivers of the trucks were Tuaregs, the kings of the desert normally seen with their famous blue clothing but here dressed in jeans and t- shirts. I finally found one driver who was taking building material through to Tamanrasset and managed to persuade him to take my bike - for a fee- through to that town where I would decide what to do next. The driver said he would meet me at the airport there in two days' time and off he and his mate went with my bike and all my worldly goods on their truck.

I collapsed, camping with the others from the blue truck that night and they kindly agreed to take me on board to the town. I spent the journey wondering how I could get hold of some replacement suspension units.

Silly me! The bike was 17 years old and though we had thoroughly overhauled the engine I hadn't given a thought to the suspension units which obviously should have been renewed for such a journey. Even German bike parts wear out!

The road to Tamanrasset was interesting as we passed magnificent rock formations standing high out of the desert, weathered into weird shapes by the sand and wind. Now that I wasn't concentrating on the road ahead to keep the bike upright I could look around me and really appreciate the beauty of this strange desert scenery. The weather was still quite cool and we were all wearing jackets while travelling. It was also advisable to keep a scarf over your head to keep the sand out of your nose.

The Tuareg tribe, the nomads of the desert, have a garment called a tagglemous, or chez. It is a strip of cotton about 7 feet long and 2 feet wide and is cleverly wound around the face and head so that only the eyes are uncovered. The men have a quick, deft way of putting these firmly on and they need very little adjustment all day when correctly donned.

When we arrived at Tamanrasset we went to the airport to meet the truck with my bike but there was no sign of it and the officials there had no knowledge of the vehicle in question. I was worried but there was nothing I could do so we went into town to collect the mail, look around and find the nearest pub.

CHAPTER SIX

Tamanrasset

Tamanrasset! Though with its sparse buildings not immediately impressive to our Western eyes, it is and has been for centuries a very important place. Originally a military outpost to guard the trans-Saharan route, it was a terminus, a centre for a network of caravan trading from Kano, Lake Chad, Agadez and Zinder. It is a link along the trading route that weaves through these areas from countries even further afield. It is still today an important junction. Although it has a hot desert climate with 40+ temperatures in the summer and cool winters, it is an oasis which grows citrus fruits apricots, dates, almonds, cereals and figs.

Because of its proximity to the impressive Hoggar Mountains, Tamanrasset is now a thriving tourist destination attracting botanists, geologists and archaeologists who study prehistoric relics and plant life in the Hoggar. The area also attracts 4WD travellers who love to pit their vehicles and daring against the desert tracks. To a lesser degree this was also true in 1974. Because there is no language barrier, the majority of European visitors are French.

The main inhabitants are the Tuaregs, the proud nomads of the desert. Their very name means 'ways or paths taken' and they travel extensively in many parts of the Sahara wearing their blue tagglemous headgear as protection against the harsh desert sun and wind. This garment is coloured blue using the indigo plant and the dye often impregnates the skin giving the wearers the name 'the blue men'. They consider themselves the kings of the desert and often use other tribes as their servants, a fact that Jacky and I became aware of during the course of our time in Tamanrasset.

While the blue truck was parked in the main street and we explored the town, a Landrover drew up and in it was the driver of the truck that had transported my bike. He instructed me to come with him to claim it. Jacky came with me in his vehicle and we were driven to a caravan situated on a building site behind the town. We entered to the sound of Beatles music playing on a cassette recorder and were greeted by two men sitting at the table, drinking wine and eating dates. The wine was in a cardboard box (a cask) and it was the first time I had seen such a thing. I only knew of it coming in bottles!

The men seated introduced themselves as Mustapha and Smiley, a geologist and a works manager of the company that was building an American/Russian laboratory there. They were Tuaregs, tall, good-looking men who spoke a little English as well as French and we spent the afternoon with them getting quite drunk on the wine. My bike was safely in their keeping and they said that Terry's truck with all the group could set up on their building site area. This was a great help as normally there was no camping allowed within 10kms of the town.

Obviously I had a problem; my bike was unrideable without suspension units and I had to find some. I knew that the police in Morocco and Algeria used BMWs and in the next few days I tried to find a source locally but was unable to do so. I decided to send a telegram to Don Dew (my tank sponsor) as I knew he would have no trouble locating some and sending them out to me. However the next international airport for delivery of the spares was in Kano, Nigeria – approximately 1000 miles away across the desert. That meant I would have to find transport for myself and the bike over this distance. This would not pose too much of a problem as there were all manner of trucks taking loads of people and produce across the desert. At least I wouldn't have to battle the sand and corrugations myself!

In the meantime, while we were all resting in the delightful oasis of this little town our new found friends were taking Jacky and me, and the attractive Houaine, plus some of the others, out on sightseeing trips in their company Landrovers. The temperature at that time of year was about 22 degrees, with blue skies and clear air providing magnificent vistas across a desert landscape shimmering with mirages.

One day we drove out to the Hoggar mountains to investigate caves with prehistoric paintings and we climbed up wind- formed formations and posed dramatically on top. Smiley and Mustapha also took us to the camp of one of the nomads, a wizened little lady who invited us into her goatskin tent to have the inevitable three glasses of mint tea. She had a huge block of sugar from which she chipped off lumps using a truck valve. We were to realize that this was the normal kitchen instrument for the job. Her wrinkled face lit up as she handed round the tea and, though almost choking on its sweetness, we dutifully drank the three servings. She showed us some bracelets made of leather and beads, very tightly sewn and I bought one which remained on my wrist for a very long time afterwards. We thanked her for her hospitality with many smiles as she did not speak French or English.

Our hosts at the building site were perfect gentlemen and even fed us at night with huge meals of meat and couscous prepared by their houseboy. However, there was one incident which occured while Jacky and I were sleeping in their caravan. I came back one evening to find that my down sleeping bag had gone missing. Of course this was a very valuable and necessary part of my equipment. I informed Mustapha and he immediately started shouting at his houseboy. The boy disappeared into the night and I had to borrow some blankets to sleep in. However, the next morning my sleeping bag was 'found'. Mustapha had threatened all the houseboys on the building site with instant dismissal if the bag was not returned. I was mighty relieved; it's good to have friends in high places.

The blue truck with the gang had to leave to keep to their itinerary so I was left to arrange my transport to Agadez, the next Saharan oasis. Jacky, bless her cotton socks, elected to stay with me. It was very nice of her and I really appreciated it.

Our Tuareg friends told us to go to one of the cafes in town where the owner had a fleet of trucks that crossed the desert regularly with loads of dates. We found him and he informed us that if we were at the customs shed at 7am the next morning with the bike there would be a truck there to take us. After rising early the next day we waited by the shed, an isolated, single-storeyed, makeshift building surrounded by desert. There was no sign of a truck. We waited until midday with no result so we went back to the owner.

"There's no hurry, this is Africa, he will come," was the response.

We returned to the shed to wait. I had a copy of *Papillon* and was content to sit and read and Jacky also had a book to pass the time. We were beginning to realise that our European way of thinking had to be changed. Things were never done on time here and one had to be relaxed about it.

By nightfall there was still no truck and the owner said, "Come again tomorrow."

In the morning he announced that he wanted £30 in advance so that the driver could get his provisions for the journey. I was frantic that my money would disappear and we would still be left sitting there.

Suddenly, seemingly out of the blue a Berliet truck appeared and stopped. It was piled high with a full load of dates in sacks. Three men appeared, took hold of the bike and hoisted it on top of the load. We climbed up as well, but as the truck took off I realised, with horror, that my hand bag, containing all my documents, was still lying on the ground. I shouted to the men around, they hurriedly threw it up to us and we were off!

With Jacky and me perched beside the bike, in the early evening, we eventually headed out into a desert of sparse scrub as far as the eye could see. We were so happy to be on the move at last that we didn't stop to think that we were two young ladies going into the middle of no-where with three unknown Arabs.

"Hey, Jacky," I said merrily," it is Saturday night and we are out on a date (truck).'

CHAPTER SEVEN

<u>Agadez</u>

The Berliet truck followed the rough track through the sand until about 7.30pm when we stopped in the still evening air for a meal. This was made from some of the goat meat that had been purchased in town and hung on the front of the truck next to water containers made of goat skin. Goats are used for many purposes as they are prevalent in the desert. Unfortunately they are a mixed blessing for, although they provide food and utensils for the nomads, they are very destructive as they pull everything up by the roots. People say that the desert is being expanded by the goats as nothing has a chance to grow when they are around.

The meal that the driver made us was surprisingly tasty and it was followed by the ritual tea drinking. The mint, sugar block, glasses and engine valve were kept in a tin box mounted on top of the cab.

While travelling along the track we had noticed the abandoned vehicles and realised that the old engine valves for sugar breaking were an easy commodity to obtain. After our meal and tea everything was repacked aboard and we were off again into the night. The driver, a middle-aged, sombre man obviously knew the desert well as the only tracks to follow were wheel marks in the sand plus there appeared to be markers of old petrol cans or tyres. Our driver made several detours but kept us moving until about 11.30pm. The three men spent the night sleeping on blankets on the ground while Jacky and I were reasonably comfortable in our sleeping bags on the date sacks.

At dawn we were off again – after the mint tea. It was becoming hotter and we were trying to cover our faces by wrapping our inner sheets around our heads. One of the men made a tagglemous up for Jacky which, as it was expertly executed, stayed in place all day even though we were atop the truck in the breeze. My own efforts were less successful.

We caught up with the blue truck at the Niger border. They had waited a day for another vehicle to come by as they were not allowed to pass the next section unaccompanied. The two trucks set off together but soon parted company as ours stopped at an army post where one of our men left. He was an army officer stationed at this outpost.

That night when the other passenger climbed up to get their sleeping blankets he came over to where Jacky was already sleeping. She woke to feel his hand on her bottom and called out to me.

As I was already awake, I called back. "Don't worry, Jacky, I'm watching him".

At this point Jacky mustered up certain French words (which she had been told to use only sparingly) and told him where to go. He called down to the driver and they had what seemed to be a pretty protracted conversation.

We thought we were going to be in trouble. We were, after all, at their mercy out here in the middle of the desert. Not exactly anywhere to run to.

I thought of the Doris Day song, *'Que sera, sera - Whatever will be will be.'*

"Just lie still and ignore him," I whispered to Jacky then closed my eyes and prayed.

The driver down below obviously thought better of any further action as he shouted up something to his friend which made him leave us and climb down. We both breathed great sighs of relief as they settled once more on the ground.

The next morning a strange thing occurred. After our mint tea we packed up ready to leave but it appeared that the battery was flat and the truck would not start. I was wondering how on earth we were going to overcome this problem. The driver and his mate didn't seem at all worried and just sat around smoking and chatting. About ten minutes later another truck appeared out of no-where, reversed up to our truck and pushed it from behind, effectively bump-starting it. Keeping the engine ticking over they brewed up more mint tea, had a short chat and then the other men jumped into their truck and went back the way they had come. Without any modern two way radio system how they knew we were stuck and our location I shall never know.

We travelled on but suddenly were engulfed in whirling sand so thick that visibility was down to nil and we had to lie flat to protect ourselves from the invasive and abrasive effects. We stopped while the sandstorm blew and as it gradually passed a group of people appeared out of the sand, mainly women and children.

It was a nomadic group of Tuareg. In their long blue robes and with their stately bearing, high cheekbones and smooth brown skin they made an impressive sight. The women had beaded and silver jewellery and some of their hair had thin braids either side of the face. The children had shy smiles and viewed us with suspicion, hiding in their mothers' robes. One young girl looked to be no more than about 14 and her pregnancy was just visible. The drivers gave the group some bread.

Some Tuaregs settle in towns like Tamanrasset and in such places make necklaces and bracelets of beads, such as the one I had bought, to sell to tourists. Our driver obviously knew this group. When we first met him in town he was wearing his blue gown with a white tagglemous and looked magnificent. The working togs he wore now weren't nearly so romantic.

Eventually that night we reached Agadez, a small town founded in the 14th century by the Tuaregs. It was another important crossroads for caravans from Kano, Timbuktu, Ghana and Tripoli on the Mediterranean shore. It was later part of the Ottoman Empire and then ruled by the French from 1900 until Niger's independence in the 1960s. Like most of the Saharan towns it was a collection of flat-roofed, mud houses and it had a camel market. There was more vegetation than previously in the desert but the town area was dry and dusty.

Agadez is most famous for its jewellery – the Agadez cross. These beautiful items are made from silver mined nearby. The local craftsmen make different models of this cross and the origins of the shape are vague. They may have come from a Christian influence such as the Coptic Christians in Eygpt or they may have been formed for geographical significance relating to the different Tuareg tribes, showing the cardinal points of their location in the desert. No-one is sure of the true meaning but they are beautiful objects made by the lost wax method, with moulds of clay. Now they are found in many top class jewellery stores in Europe but of course the cheapest place to buy is Agadez- just rather a long way to go. I didn't purchase one at this time, not being aware of them in the short time we were there but I did buy one, years later, when back in Morocco.

While in Agadez, we had to clear customs, having entered Niger. Our driver took us to where he was offloading but it was too late to take off the bike and we were told to come back in the morning. Two young Tuareg boys, whose family lived in a tent, kindly took us to their home and we slept under a shelter in their back yard. Jacky put everything she was frightened of losing in her sleeping bag with her.

CHAPTER EIGHT

On Through the Desert

The next morning there was a crowd of men and boys around to help with unloading the truck. On seeing European faces they were all hustling to handle my bike, loudly demanding money to do so. As I had already paid a large amount for the transport I was unhappy but as they became quite aggressive I gave out my last bit of change.

There remained the problem of covering the next 250 miles to Zinder. Asking around we eventually located another truck that was going that way and, after parting with another £20, the bike was loaded on the back of the open vehicle, under my supervision, and we were herded, like cattle with about twenty other people; men, women and children. They sported a variety of clothing from Western jeans and t-shirts to jelabas and tagglemouses and chez. Large brown eyes viewed us with curiosity- two white women unaccompanied by men, both looking decidedly tatty after our travels atop the date truck.

The truck eventually started its journey in the late afternoon. The driver drove like a maniac, flat out, and people were being sick all over the place. Normally I would have been too and certainly didn't feel at all well but was kept busy trying to keep the bike upright as the ropes were working loose and it continually fell over. I was not only worried about the bike, it felt like every bone in my body may be broken too. The only relief was when we stopped to let some of the passengers off along the track and pick up others and then, to our huge relief, there was about an hour's break in a small village.

By this time it was dark and the scene was very picturesque. The villagers were selling their wares, mainly a variety of food both sweet and savoury and small children illuminated the stall with their little tilly lamps. We bought some water to re-fill our canteens and regained our place in the truck early, mainly to get away from the constant hassle of children asking for money. We continued our journey, once more at breakneck speed and were unable to get a wink of sleep all night. We stopped twice when the truck became bogged in deep sand and all the passengers dismounted to dig around the wheels with our hands.

About sunrise the driver stopped and we slept for a couple of hours before driving the final few kilometres into Zinder. The rosy glow of dawn illuminated the sand-hills and we caught glimpses of camels ridden in the distance. The truck came to a halt in the middle of the market and the other passengers soon disappeared into the throng leaving Jacky and me to be tormented by crowds of men screaming at us for money to unload the bike. It was extremely hot and I just wanted to get the bike going and get as far away as possible from the crowds who were pressing all around us. I had no more change to give out but finally, with the aid of the driver we got the bike safely on the ground.

As the road from Zinder to Kano was sealed, the idea was to ride the bike carefully the last hundred miles to the Nigerian city. Amazingly, after the tremendous bashing it had taken, the bike started and I breathed a huge sigh of relief. Jacky and I mounted and I rode tentatively through the crowd who were still screaming at us for money. We made about 10 feet when the drive stopped functioning and we came to a halt. With tears of frustration in our eyes we pushed the bike out of the market place and onto the road.

"What's wrong?" said Jacky, anxiously.

"I don't know, it's something to do with the drive shaft," I answered miserably.

On inspection I found that the four bolts connecting the gearbox to the drive shaft had disappeared. I thought they may have disintegrated given the tremendous pounding that the bike had taken. There was nothing we could do but leave the bike in a garage forecourt and go and find the nearest pub while we discussed our next move.

"Well, there's one good thing," said Jacky, brightly. "Terry's truck is still behind us, we saw it briefly in Agadez."

"You're right, Jacky, maybe they can help."

We had arranged to leave a note for them at the Post Office but as it was the weekend they wouldn't be able to pick it up so we decided to wait for them in the hotel , which was, as luck would have it, situated opposite the P.O. It was a beautiful old colonial-style

INTO AFRICA – WITH A SMILE

hotel, albeit somewhat shabby. We sat for a while, relaxing in the courtyard, taking in the scene and were introduced to the delights of citreon presses. These were pure juice, squeezed fresh from local lemons, topped up with lemonade and served with lots of ice. They were delicious but expensive so we bought only one as we had very little local money. We then asked for a jug of iced water and kept watering it down.

Here, in the southern Sahara region the physical appearance of the people had changed dramatically. The angular lines of the Arab features were replaced by soft, round Negroid faces and the dress was more colourful; the ladies in their wrap round cotton dresses and the men in long flowing robes and small beaded skull caps.

Our next challenge was to find somewhere to sleep that night. While we were sitting in the beer garden under the shade of a table umbrella, we were approached by the local police. It appeared that we had to go through some form of immigration as we had crossed the border from Algeria. Also in this town it was necessary to check in and out if not just passing through. So, off we went to the police station to complete the formalities and whilst there asked about a hostel as we had already seen some young American social workers around. We were directed to the 'Hostel de Jeunes' which, in this case, seemed to be some sort of youth club. We pushed the bike inside and it was indicated that we could sleep on a filthy mattress on the floor for the equivalent of 5/- a night. There were no toilet facilities as they were locked! However, I was happy that the bike was off the road as it had all my worldly goods strapped to it and the villagers still weren't happy that we hadn't given them any money. We returned to the pub which was the only attractive place in this dry, dusty town and the beer garden provided the only shade.

During the afternoon some of the American Peace Corp workers came there to relax and we asked them about this organisation. It is a volunteer program run by the US government to provide people outside the States with technical assistance and also with the aim of helping these people understand US culture and also those in the US to understand the cultures of other countries. It was started by President Kennedy in 1961. He saw it as a means of countering the stereotype of the 'Ugly American' and Yankee imperialism especially in the emerging nations of Africa and Asia. The organisation began recruiting in July 1962 with popular comedian Bob Hope as a front man recording radio and TV announcements to attract participants. Those selected had to be American citizens, typically young adults with a college degree willing to work for two years abroad after three months' training. They worked with foreign governments or NGOs in schools, with non-profit organisations and entrepreneurs in education, info technology, agriculture and environment and in hunger relief. After two years they could apply for an extension or go home. The Peace Corp image had its ups and downs both in America and its recipient countries and at that time, in 1974, under President Richard Nixon, was not getting all the support that it needed.

However, the participants we met were enjoying their life outside the US, here teaching in the local school, and were happy to chat with us. As the bank was closed they kindly exchanged some of their Nigerian money for our American dollars so that we could continue drinking. The Peace Corp guys said that we could stay at their hostel but as we had already organised our spot at the other place we declined. However, when we returned to our hostel we found it swarming with kids who had jumped on the bike breaking the side-stand and were running in and out of our room screaming and shouting. They wouldn't leave us alone even when we threw things at them. In the end I walked up to the Peace Corp hostel and enlisted the aid of two of the boys to help push the bike up to their place.

Just then the blue truck arrived with Terry and his merry crowd and we arranged to meet them in the pub when they had set up camp. Jacky and I stayed at the hostel and had a very welcome shower. It was the first time our bodies had seen water for several days and it sure felt good. In the wind and sand our hair had become tangled and matted and it took some time to get it clean and be able to run a comb through it.

I spoke to Terry and the passengers about the possibility of me and the bike hitching a ride with them. They agreed to let me squeeze the bike in between the seats and me on the end. They were very good, considering the discomfort it would cause but at least it was a smooth road and only for a couple of hours. I was very lucky as I couldn't face dealing with yet another truck driver and the demanding crowds to load and unload the bike.

After completing yet more formalities to leave town we bade farewell to the helpful Peace Corp people and drove out of Zinder onto the blessed bitumen. Along the road there were more villages and increased vegetation. We noticed a number of anthills and many goats. Also the traffic became heavier as we neared the border. Crossing from Niger to Nigeria we finally drove into the city. It wasn't really that big but, after the places we'd been, it appeared huge. It was certainly a very dirty place; open sewers, broken cars and rusting farm implements. There were many cyclists and small stalls selling brightly coloured drinks. Beef cattle were being loaded onto huge lorries. However, despite its run-down appearance, we were glad to be there.

Brick works in Salah

Entering the desert

Saharan tea house

The road to Tamanrasset

Jacky, Tamanrasset

Truck passengers in Hoggar Mountains

Houaine, Linda, Mustapha

Hoggar ladies hut

Bike on the date truck, Sahara

Above and below,
Sahara from the date truck

Jacky on the date truck wearing a tagglemous

Desert chat

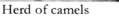
Herd of camels

CHAPTER NINE

Kano, Nigeria

Kano lies in the north of Nigeria and until independence in 1963 it had been under British rule and was the peanut capital. Now it was experiencing an oil boom and there were workers from many different countries living there so the service industry was also expanding. However, it appeared that many white people who owned their own business were trying to sell up and leave the country before their businesses were taken over by the Africans.

I camped behind the Central Hotel, where the truck parked. This area was the main meeting place for Overlanders. Meanwhile, Jacky, Terry and another truck passenger, Lesley, found themselves a cheap room in a motel.

Because of its links with the UK, Nigeria is an English-speaking country and this was a huge relief to people like myself who really battle with French. Although I learnt it at school I think my final exam result was 30% and apart from a few short trips to France I hadn't had much chance to use it. I was therefore pushed to converse more than very basically and had been struggling , though improving a little, in the last few weeks in the French-speaking countries. Now it was back to English again and the locals at the campsite were soon hassling us to change money. As the black market rate was much higher than the banks we were happy to deal with them. There was also a self-appointed Nigerian camp-site mechanic named Gregory who had already decided he was the man to help me fix my bike.

However, that night all we wanted was a beer so, after cleaning up and donning our best clothes we were off to the Kano Club. Terry warned us that this establishment did not like scruffy Overlanders so we entered in twos and threes using the pretence that we were visiting Kano on business or had flown in on holiday.

The club catered mainly for the Europeans who lived and worked there although there were a large number of Nigerian members from the upper echelons. There were all sorts of sporting facilities including tennis courts and a large swimming pool, a necessity in the 30 degree heat. The restaurant served good, cheap meals and the Star lager was also very affordable and welcome. It was like paradise to us – the hardships of the desert almost forgotten. All the residents homed in on the newcomers and we were bought food and drink all night and, it transpired, taken out and shown around the town the whole time we were there. Houaine, particularly, back in her element sporting her good clothes and make-up, was especially in demand.

There seemed to be an attitude prevalent at that time in this club, that social status was enhanced for African men if they were seen with a white woman. It became very difficult to be in the bar without being approached. One evening Lesley, Terry and Jacky were having dinner and during the meal Terry had to excuse himself. He had barely left the room when a note arrived inviting Lesley and Jacky to join some other male diners. They politely declined.

What a shame to marginalise black women by elevating white women.

A group of Nigerian pilots, who flew Russian Migs, fancied their chances with Houaine and one invited a few of us to dinner at his upmarket apartment. The meal was chicken, rice and vegetables and was cooked by his houseboy. When it arrived on our plates we were all looking forward to a delicious treat but when we raised our forks toward our mouths to take a bite were overcome by the strong chilli fumes and, eyes watering and tongues burning, were unable to eat it. Our host was furious with his houseboy and screamed at him. The chicken was taken away and we had to make do with the lesser flavoured veggies and rice and accept the profuse apologies of our host who was mortified. I don't think the houseboy had a job the next day.

Chilli is added to almost all food in Central Africa, probably acting as a preservative or to cover the flavour of not- so- fresh food. Africans obviously have a higher tolerance to its burning properties. I often bought locally-prepared food or cooked my own. One day I bought chicken fried with chilly at a roadside stall and, thinking to save my mouth from being burnt, was stripping the skin off the outside when a piece flew up into my eye. I thought I was going to be blinded for life, leaping about the street covering my watering eye and yelling. Finally it cleared and I could see again. I was even more careful in future.

INTO AFRICA – WITH A SMILE

Another chilli incident was when the blue-truck passengers first arrived in this region and cooked their usual communal meals. The two cooks rostered on had been shopping in the market and, not recognising the red chillies for what they were, bought a kilo and put them all in the stew for the evening meal.....it was inedible and the whole pot-load had to be thrown away. The group went hungry that night and the cooks were in disgrace.

The morning after our first foray into the Kano Club I was up early, despite my hangover, to start working on the bike. Gregory, young and keen, was there at my side, totally amazed that a woman should even know how to hold a spanner let alone strip an engine. We found that the four bolts from the drive shaft had slid down the casing to the bottom. We managed to fish them out with a little piece of wire and a lot of patience but found that the threads where they fitted into the gearbox had stripped. Gregory took my gearbox to a workshop, me trotting alongside, and there the engineers re- tapped the threads and found four more bolts to fit. Gregory charged me about £4 for his work which at the time I thought extortionate but he had spent the whole day helping me and the job was well done. We had also taken off the suspension units ready to fit the replacements when they arrived.

In our campsite there were other tourists and travellers: a New Zealand couple in a Landrover and another young lad on a 350 Yamaha trail bike. They were heading up north the way we had just come. There was another Australian couple in a similar vehicle also going north and, while chatting to them, I found that we had belonged to the same motorcycle club in Sydney a few years previously. On checking dates it transpired that we had even been to the same New Year's Eve party in Sydney in 1970! Small world. Although we hadn't actually met before we had many mutual friends so there was a bit of news swapping.

On the second day there, an Englishman arrived on a 175 Honda. He was a small, wiry, dark- haired fellow with an Indian look about him. His name was Kenneth Tilley and he was on his way to Rhodesia. He had driven down from UK in a car, across the desert and it was now clapped out and, having managed, somehow, to get it through without a carnet he had sold it here in Kano and bought this little bike to continue his journey south. I had had enough of travelling with the truck and was sure they were fed up with me, so I

43

suggested to Ken that we teamed up to continue together as it would be much easier travelling with a similar vehicle. That was fine with him and he offered to help me fit my new suspension units when they arrived.

We rode out to the airport on his bike and fortunately the parts were there and, after fitting them, the bike was rideable. I just had to get some welding done on the side-stand and silencer brackets. Altogether I spent four days working on the bike but every day, come 4pm, I downed tools and went for a swim in the pool at the club and every night went out to drink with the residents.

One night I met a young, fair-haired teacher, Eoin, from Ireland. As I knew many Irish songs we got into the beer and sang them all. It gave me an opportunity to bring out my lagerphone, which I had been carrying in a bag on the bike. For those who don't know what a lagerphone is – it is a broomstick with (lager) bottle tops attached with nails. It is played by hitting the broomstick with a corrugated 'handle' of wood to make it rattle thus giving a percussive beat. Mine was made collapsible to fit on the bike.

Eoin came from Drogheda, a town that had been ransacked by Oliver Cromwell and therefore he was quite happy to sing my entire repertoire of Irish rebel songs. We sang all night and in the morning swapped addresses and I promised to visit him when he had finished his year's teaching in Africa and I too had returned to Britain.

Another Celt I met at the Kano Club was a Scottish transport manager who told me to bring the bike around to his workshops and his mechanics would weld up the side-stand and silencer. They did a great job as the welds lasted the rest of the trip.

Ken and I left the next day. Little did I know what I was letting myself in for with my new partner!

Chad to Zaire

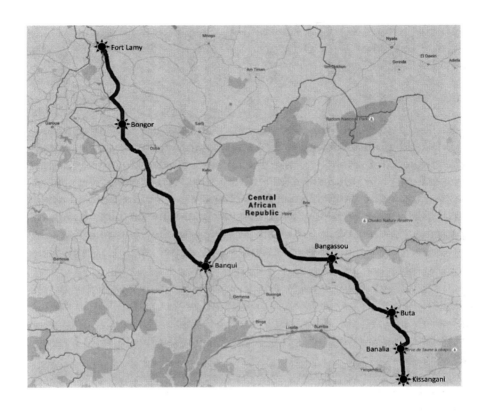

CHAPTER TEN

On to Chad

Ken and I left early the next morning and, as the road was sealed, completed almost 300 miles. There were many people walking along the route and, now out of the desert region there was more vegetation. The population was mainly the Hausa tribe, Negroid in appearance. The women wore bright, wrap- around dresses and head-gear. They ambled along with baskets of goods on their heads or carrying children wrapped tightly on their backs. The men, also in long garments had coloured skull caps rather than the previous Tuareg turbans.

There were many road-works that slowed our progress but it was an easy ride with no sand or corrugations and my bike felt fine with its new suspension units. Ken and I made camp by a well and watched the bullock-drawn wagons come to load up water in ceramic pots.

We arrived at Madaguiri the next day, another scruffy, dusty town and went to the market to get Ken's tyres changed. It was easy to find replacements as there were hundreds of 175 Hondas in Nigeria and a good percentage of them were in various stages of disrepair in open-air workshops. We selected a couple of tyres with more tread than Ken's and had them cheaply fitted.

Making camp with a shady spot next to a small zoo, we met up again with the blue truck. The gang had already set up and were out exploring. On joining them for a drink we met the European zoo owner and had a fascinating few hours with him as he explained the variety of animals.

Unfortunately my stomach was beginning to play up, ironically just as I had started purifying the water. We had no space to carry water on the bikes so had bought it at local wells in the evening when we stopped. It was usually pretty muddy so I added water purifying tablets which gave a horrible chlorine taste. The liquid was also so gritty that you could just about chew it but at least we could make a cuppa on my Optimus stove.

Next day the dirt road re-appeared and it proved to be about 100 miles of very bad sand to the border of Nigeria and Cameroun. It was a hot day and Ken, unused to riding a bike, let alone in sand, kept falling off. When he didn't, his load did and I spent nearly all day stopping to pick him and his bike up. It was exhausting, especially as I had the runs by then and wasn't feeling my best. Erecting the tent at the roadside I just collapsed without bothering with food.

There was more dirt road although it improved a little and we crossed a small ferry into Fort Lamy in Chad. Ken didn't have a visa for Chad but we managed to get in although immigration held our passports for about three hours while they considered his entry. While we were waiting, I asked Ken why he hadn't obtained the necessary paperwork i.e. visas and carnet before he left UK and he told me his amazing story.

Apparently Ken was Rhodesian and he and his girlfriend, Deirdre, had decided they wanted to drive back there where they had friends and Ken could continue his job in the prison service. They owned a normal saloon car and didn't even consider that they may need things like visas, inoculations or a carnet. After all they had British passports and wasn't that sufficient? To cap it all Deirdre was several months pregnant. This proved to be an advantage, at least for a while as, when they began to be held up at borders because of their lack of paperwork Deirdre would start moaning, clasping her expanding tummy and Ken would roll his eyes and say, "Oh my God, the baby is coming", and the officials, uncertain about 'women's business', would wave them on.

However, after a very uncomfortable trip across the Sahara in sand and corrugations, Deirdre had had enough and flew back to the UK. Ken sold the carnet-less car in Kano, bought the cheap bike and, with only a small amount of money to sustain him, was relying on Deirdre to send him money en route. Now with this African vehicle he may have less trouble over lack of carnet but he would certainly need visas and inoculation certificates and I wondered how he would fare at the many borders we had left to cross. I would soon find out!

Fort Lamy, now known as N'Djamena, is the capital of Chad and situated on the confluence of the Chad and Logone rivers. Having the advantage of river transport it became a major trading city and had gained recent fame as one of the locations of the French Foreign Legion. This military force is unique, created exclusively for foreign nationals willing to serve in the French Armed Forces. It has been romanticised in song especially one sung by Frank Sinatra which intimates that a broken-hearted young man could do nothing better than sacrifice himself by going off to fight anywhere the Legion would send him. In fact to join the French Foreign Legion one has to undergo strict physical and psychological training. Because it is made up of men from diverse countries, with no feeling of patriotism, the legion encourages an 'Esprit de Corps' to inspire loyalty and high morale. One benefit offered is that any soldier wounded in its service can immediately become a French citizen. The force is maintained for rapid intervention when guarding French interests in North Africa, many countries of which had lately been under French rule and still have French-controlled businesses. Such was the case in Chad. There had been a conflict with Libya from 1969-72, therefore the FFL had been stationed at Fort Lamy.

As the capital and a busy business centre, Fort Lamy was urbanised and very expensive. A coke cost 10/- and citron presses likewise. The same for a ham roll. We were so hungry and thirsty that we had some but, I admit, we didn't get the bill until after we had eaten so maybe wouldn't have been so hungry if we had known the price beforehand.

A piece of waste ground by a hotel was our camping spot for the night and I left a note for Jacky at the P.O. We passed a couple of overland Bedford trucks going the other way and , judging by the state of the vehicles, the roads we were about to face looked bad. Sure enough after another 100miles of bitumen we were again on dirt.

Stopping at Bongor for petrol we were informed that there wasn't any in the whole of Chad unless it could be obtained on the black market. Since the blue truck still had 10 gallons of my fuel, the best idea was to stay put until it arrived. There wasn't much option. Off we went to get a drink, always necessary in the hot climate, and met a group of Frenchmen who were teaching in a small school nearby. One of them, Anton, in his charming French accent, invited us to stay at his house and we spent a very pleasant two days with him sleeping on rattan beds under a mosquito net and enjoying the colonial life until again connected with my fuel on the truck.

Into Africa – With a Smile

After one more luxurious night with the lovely Anton we rode like bats out of hell to catch up with the truck. Despite the dirt road we did over 250 miles that day getting to the exit border of Chad that day only to find that the truck had not yet arrived.

I was in real agony by this time as dysentery had struck in full force. I was doubled up with pain and even the border guards were concerned for me. But we pushed on over the border out of Chad and were looking for somewhere to camp when there was an almighty thunderstorm with torrential rain. We ran for cover under the spreading trees around nearby native huts. It was a moment I shall never forget. With my bowels exploding, I squatted under the fruit trees in the pouring rain, mangos on the ground around me, not caring about the curious eyes of the villagers; I was in such agony.

When the rain stopped and with my trousers back in place, we asked the natives if we could pitch our tent there. They were friendly and, obviously taking pity on me, even made a fire for us. After a cup of black tea I was sick and felt a lot better. The natives sat around watching but didn't interfere and I think we were the evening's entertainment. I must say that every time I see a mango I relive this scene, but I still like them.

CHAPTER ELEVEN

<u>Central Africa Republic</u>

When we arrived at the Central African Republic border Ken was refused entry as he didn't have a visa (surprise!) the guards said that he would have to go back to the embassy in Fort Lamy to get one. (500 miles away.)

I begged and pleaded saying it was dangerous for me to ride alone and that we would obtain the necessary paperwork as soon as we arrived in Bangui. I said we didn't have enough fuel to return all the way to Fort Lamy but it was all to no avail. We decided that I should carry on and wait in the village that we knew, according to our map, was a little further up the road. Ken would see what he could do about chatting up the guards when I left and he would meet me there in an hour. I must say he was a very positive thinker! So I continued through the border, stopped about half a mile away and sat under a mango tree surrounded by curious kids. Ken eventually walked through and met me. The guards had relented enough to allow him to come through to buy food for his return journey and had even sold him some petrol but they were adamant that he could not enter CAR with his bike without a visa.

The only option was for him to evade the border post by riding back out of their sight and entering the thick jungle which was now on both sides of the road. He would then make his way through to the other side. So I rode a couple of miles further up the road and yet again sat and waited. Amazingly, after about an hour and a half Ken appeared on his bike looking decidedly bush-wacked. He had to shin up a tree at one stage to get his bearings as the undergrowth was so thick, but he had succeeded!

Just as he arrived the blue truck rolled up. They had gone off the main track to a small village to obtain water and we had bypassed them in CAR.

Ken and I went on ahead until his chain broke and we all camped by the side of the road using some abandoned huts as shelter. During the night we heard weird wailing noises which frightened some of the group. It sounded like a mad person and I understand that in some parts the locals believe it is the spirit of a god and is a bad omen, maybe like the wailing of a banshee in Ireland is thought to be the sign of imminent death. I learnt later that it came from a very small and harmless animal but I don't know its name.

After mending the Honda's chain we set off after the truck and were doing well until Ken hit a dog and fell off. While I was bandaging him up, a local made off with my gloves but, luckily, nothing else. We were running low on fuel and just made it into the next town but there was no sign of any fuel there. I had used up the fuel I had on the truck, sharing it with Ken, so we had no option but to stay in town and ask around and also find someone to change money. A European obliged us with the money and, after waiting four hours at a petrol station for it to open we eventually obtained some fuel. We literally had to fight for it. As there is no such thing as a queue in Africa, people were pushing and shoving. Everyone in town wanted fuel and I was afraid that the supply would run out before we obtained any. Ken said it was quite a sight watching me, a diminutive white woman determinedly pushing away big black men to grab the fuel nozzle.

The blue truck had been and gone before us. Perilously low on fuel, they had hoped to get to the next town but we came across them on the side of the road just a few miles short of it. However they were being helped by a couple of young Welsh men travelling in a Combi van, John Morgan and Mike O'Connor. They were cheerful, lively characters, John with dark hair and a wide grin and Mike , the taller of the two, fair and equally attractive. They were taking two of the truck's passengers armed with jerry cans to the settlement and, it transpired, were to be involved with both the blue truck and myself throughout the African journey.

While we were stopped there and Ken was talking to the others I looked at my Michelin map and it indicated that there was a waterfall nearby so I went for a walk in the bush to look for it. However, I became hopelessly lost and spent half an hour walking round in circles in the tall jungle, panicking and screaming 'Ken!' at the top of my voice. He finally heard me and kept shouting a reply so I had an indication of the way back to the main track. It's very easy to get lost in the jungle.

There was only fifty miles to cover to Bangui and we arrived there filthy dirty and only too happy to camp by the Rock Hotel next to the Ubangui river. The blue truck soon arrived and they set up camp in this area. We all immediately threw ourselves into the water despite previous warnings about bilharzia.

This is a disease that is spread by parasites released from fresh water snails and these are mainly found in lakes or dams. It is pretty nasty and very widespread in Africa, affecting a large percentage of the population. It infects the urinary tract or intestines causing a variety of symptoms including fever, fatigue and diarrhoea and blood in the urine. Early warning is a mild itching and dermatitis of the limbs, especially the feet as they are more likely to come into contact with the snails. Of course, in the heat, most people, especially children, throw themselves into any available water and we were no exception. We needn't have worried though, as the Ubangui River is known for its rapids and it was very unlikely that we would have caught the disease in its fast-flowing water.

We all made ourselves comfortable in our new home in this capital city and yet again prepared ourselves to indulge in the luxury of urban surroundings. This city seemed to be very well organised with modern buildings, hospitals and libraries. The houses were clean with colourful gardens. It seemed a good place to relax but in my case the stay here proved to be very eventful.

CHAPTER TWELVE

Bangui

The Central African Republic, of which Bangui was the capital, was previously under French rule but they were forced to grant independence in 1960. However, the French still had their eyes on uranium deposits in the country and abundant big game hunting grounds near the Sudan border so were trying to be lenient toward the outrageous excesses of its president of the time, Jean Bedel Bokassa. He was making himself increasingly unpopular with French and other European banks due to his opulent lifestyle and overspending of government money.

In 1970 he built the first university in CAR in Bangui and in 1971 an international airport was followed by two very flash hotels, one of which we were camped behind.

Apparently, Bokassa had little tolerance for anyone who opposed him and often had them clubbed to death. In 1974 the body of his mistress, Brigette Miroux was found in one of the grand hotel rooms and it was rumoured that he had engineered her demise even if not committing the act himself.

We were unaware of these facts but I did have reason to note his extravagant lifestyle. I had noticed that the police were riding BMWs and when I had my first puncture of the trip, being lazy about attempting to fix it myself, I went along to their workshops to obtain some assistance. I was amazed to see about 200 motorcycles there, half of them brand new BMW 750s. Apparently the only time more than just a few of them were being used was when the President wanted them for an entourage while making a tour of the town. Thousands of pounds worth of machinery just sitting there! It made my heart bleed to see those lovely motorbikes going to waste, especially as I was battling through with an 17-year-old! The police were friendly and only too happy to help change the inner-tube and repair the old one.

53

LINDA BOOTHERSTONE

Our neighbouring hotel, The Rock, a large, modern concrete building, was far too expensive for any of us to dine in and we found much more affordable food in the early morning local market where we procured delicious peanut butter sandwiches and aromatic coffee. As Ken had no camping gear of his own, I allowed him to sleep in my tent as it was a good size. We were both small people and getting thinner all the time. I would have preferred my own space but we made do. When we went out during the day our gear was left in the tent, which was not locked, of course so we were at the mercy of any unscrupulous people.

The bank of the river, where we were situated, was a popular place and there were many people coming and going all the time. One day we came back to find that the tent had been ransacked and my clothes were missing. Funnily enough Ken's Pentax SLR camera was still there but my set of leathers and waterproof Barbour trousers were gone and also my spare T-shirts and underwear. Fortunately, riding out on the bike I was wearing my Barbour jacket, helmet and gloves. I now had just the clothes I stood up in. Luckily my sleeping bag was still there. As I had not had to wear the leathers since the cold in the north it was a blessing in disguise as I now had room for another jerry can.

Another day I very nearly lost something far more important. I was doing some sort of work on the bike and had taken off the tank bag which was now lying on the ground outside the tent. Suddenly a car load of young men came speeding close by and ran over my tank bag. I hurriedly picked it up to find that the stitching had been split and I would need to get it repaired. The car had done a U turn on the bank and was heading back our way. To my amazement Ken grabbed my tank bag and ran toward the car, waving the bag and shouting at them to get it fixed. They had slowed down at his approach but laughed at him and revved up to take off again. I watched helplessly as Ken, still clutching my bag, threw himself onto the bonnet as it took off up the bank and disappeared onto the road and into heavy traffic. I was speechless!! Not only had Ken put himself in danger but, more importantly, the secret compartment in my tank-bag held my carnet and other valuable documents without which I could not travel. What on earth was he doing?!

There was nothing I could do but hope for the best. After a couple of hours there was no sign of him and I began to get worried.

I phoned the hospital in case he had been admitted; I was imagining all sorts of fates that may have befallen him. Finally late that night, a bedraggled Ken came limping into the campsite, clutching my bag – thank goodness! - and with a triumphant smile on his face.

"I showed the bastards!" he said.

Apparently his ride on the bonnet of the car had been quite exciting. The driver had tried all sorts of turns and sharp braking but he clung onto the wing mirrors and they couldn't shake him off. Finally the police had stopped them and Ken, being the white foreigner, had shouted and remonstrated and insisted that the culprits be made to repair my bag. With the police escort they had all gone to a local leather worker and the bag had been duly repaired and paid for by the mad drivers. Ken had smugly made his way back to camp.

Boy! Was I relieved; my papers were still all there.

Another Ken incident followed. We were cooking dinner one night in a shelter on the beach. The wet season was approaching and the now regular afternoon thunderstorms and downpours had washed us out a few times and we had set up a bamboo structure for more protection. I was remonstrating with Ken about something, I can't remember what exactly. I was getting increasingly annoyed with him over a variety of things. One big problem was that Deirdre, having promised to send him money from the UK had not done so and he was relying on me to keep him in petrol and food. My own resources weren't limitless and although I appreciated his company on the road, his self-made problems were becoming overwhelming. Because of his lack of paperwork, every time we came to a police checkpoint he had to talk his way through, usually being so chatty that they didn't notice his lack of visa. However we still had a long way to go in CAR and his papers would also be checked going out of the country and into the next. He needed a visa for Zaire.

Ken had come back to camp one day with the happy news that he had been in a bar and found himself talking to a friend of the immigration minister.

"All we have to do is for you to go out with him one evening and he promised me he would sort out my visas."

"What?!"

"Please Linda, it's just for a drink; these fellas love to be seen out with a white woman."

What could I do? Fortunately this was before my clothes had

been stolen and I had a little dress that I could use for dress-up occasions.

The smiling, corpulent immigration official took me to dinner at an expensive restaurant and I managed to act as if I found him very interesting but put on my upper-class English-lady, shocked act when he suggested we might like to continue the evening at his place. I had to dodge his groping hands a few times but fortunately made it back safely to camp after sweetly reminding him of his obligations regarding Ken's visas. Fortunately I had titillated his ego sufficiently for the visa to be duly issued when Ken presented his passport at the office the next day. He could now officially leave CAR and enter Zaire.

So, this evening, at the camp-site, I really did have reason to be a bit fed up with Ken and was making my displeasure felt when he, in the act of cutting up some veggies, slammed the knife point into the table, shouting, "For goodness sake, stop going on at me!"

Unfortunately his hand slipped off the handle and down the blade, resulting in a very deep cut. Blood pouring everywhere I quickly grabbed a not very clean tea towel to staunch the flow and marched him off to the local clinic where it had to be stitched. This meant that he was unable to ride the bike for another seven days while it healed.

In the meantime, by the river I met a couple of young men, a French Canadian and a Japanese who had bought a dug-out for about £10 and were intending to paddle it down the Ubangui River. Apart from the rapids, hippos and crocodiles they were likely to face all sorts of other hazards and I didn't envy them one bit.

Having heard that there was no fuel in Zaire it was advisable for me to carry my own and Ken's, so with a pannier now empty of any clothes, I was able to fit in two containers carrying an extra four gallons. With my big tank and these containers we had a range of about 500 miles. Finally, about four days after the blue truck had left, we continued on the piste to Bangassou, to cross the river into Zaire, the infamous Belgian Congo. Now the surrounding vegetation was very green and lush, becoming more tropical, and alongside the road there were anthills, about 2' high in the shape of giant mushrooms.

The road was hard going in places and it was on this stretch while riding ahead of Ken that I had my incident with the man with the panga- described in the preface. Further up, by the roadside, busy

INTO AFRICA – WITH A SMILE

with tools, were a young couple; an Englishman with his Aussie wife and 6-month-old baby who had travelled from Kano in a Landrover. They were towing a fully- laden trailer and it was this they were trying to mend as it had succumbed to the punishing road.

Ken and I carried on our slow progress, passing a spectacular waterfall where, it was reported, that a girl from one of the Overland trucks had accidently fallen over and drowned a couple of weeks before. There were two main falls of water with pools above and below and, apparently, a Swiss girl had been bathing in one above when she came too close to the edge and was carried over by the fast moving water. The local fishermen were still looking for her body. This incident would have been a nightmare for the expedition leaders!!

Linda Bootherstone

Zaire to Rwanda

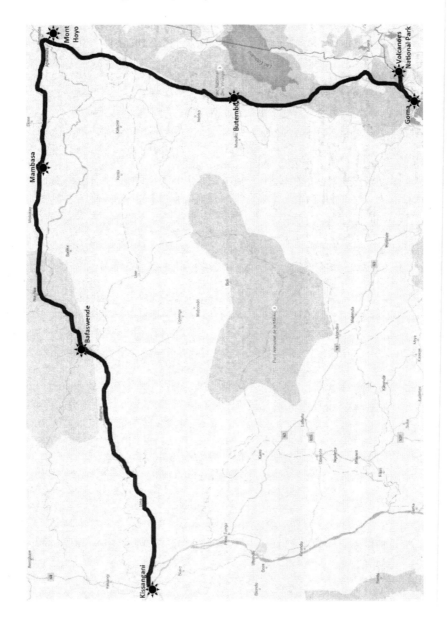

CHAPTER THIRTEEN

Into the Congo

Finally, bursting with expectations and trepidation, we reached Bangassou and the ferry crossing into Zaire (the Congo). Here the Ubangui River was about 400yards wide and slow-moving but we met with a big snag. The ferry was kept on the Zaire side of the river and used two 12-volt car batteries to power it. Car owners just needed to find one more battery, by waiting for another vehicle to arrive. Then the two batteries were paddled across in a dug-out to the waiting ferry. The Canadian mission on the CAR side had an old battery that they usually lent to solo travellers but that was only one and Ken and I with just six volt batteries in our bikes needed to find another. After spending a frustrating and useless day in unsuccessful search, we resigned ourselves to doing the African thing and just sit and wait. Eventually another traveller would arrive.

Like London buses, the next day a crowd of vehicles came altogether. The Landrover with the trailer, then another with an American couple we'd met in Bangui and finally two Volkswagen vans, one with John and Mike, the two Welsh boys. So our battery problems were solved but we all had to wait another day. The customs post was closed as it was Sunday!

First thing the next morning two batteries were sculled over to the waiting ferry which chugged back and we boarded when it docked. Driving on and off the ferry was by means of two rather unstable tree logs and disembarking on the Congo side was a bit of a drama for me. I had inadvertently left the bike in gear on the centre stand thus, when I started it and pushed it forward off the stand, it went careering off and nearly landed in the river with me and half a dozen locals hanging on but I made it safely to dry land.

At the Zaire customs post another problem appeared when the officials would not recognise the slip of paper that Ken had as a carnet for his bike. All the previous countries had passed this document as, at the Nigerian border, the officials had signed and stamped it, but the Zaire officers said it was invalid and he would have to put up a bond for his bike. This was not an option as Ken was completely broke. Now without my party dress, I didn't think that my charms would persuade them either. They insisted that he had to go back so we all put on a big, dramatic farewell act and continued a few miles up the road. Given his past performances I was sure that he would somehow manage to talk his way in.

The minute that the Ubangi River is crossed one is plunged into dense jungle and there are plenty of little monkeys around. The road is just a mud track baked hard by the sun in the dry season and a complete mud bath in the wet. At this stage the track was still dry but it was very narrow and full of pot holes and there were a great deal of buzzing, stinging insects about. Adding to this discomfort the humidity was overwhelming. We all slowly travelled about 20 miles up the road and stopped. First mistake; we shouldn't have gone so far.

After about an hour I decided to turn back and see what had happened to my troublesome travelling companion. I was very nervous riding alone as there were some especially hazardous bridges made from logs, squared off and laid lengthways from one bank to the other. I just had to aim and balance, my eyes glued to the far side. If the wheels went off track I was lost forever in the stream below. Even dropping the bike on the track would have stopped me as I was carrying so much fuel that I would not have been able to lift the bike up alone. However, I had only completed a few heart-stopping miles when I saw Ken coming the other way grinning all over his face. Joining the others we continued to the immigration post another 40 miles down the track, over more log bridges, and there, exhausted, camped for the night. There Ken laughingly related his story.

Apparently, after we left he just lay down on the step outside the customs post. When they asked him what he was doing he said that, since they would not let him through and he couldn't go back, he would just lay there until he died. When, after about an hour they began to realise that he meant it, especially as he said he would rather set fire to the bike than let them have it, they eventually became sick of him, issued temporary import papers and told him to get the hell out of there.

"Well, at least we won't have any problems with the Zaire

INTO AFRICA – WITH A SMILE

immigration, as I risked my honour in Bangui to get you a visa." I said – too soon as it transpired.

This time the problem was that Ken hadn't had a cholera inoculation, an obligation for entering this country. With a distraction of the mix of French and English we managed to get them confused making all sorts of excuses: the doctor's certificate he had had got wet and lost in a storm, that it had been stolen with his papers, set fire to and anything else we could think of. Finally, after an hour's delay they let him through. Ken was lucky- we had previously met an Englishman who, at this point, had to return to Bangui to get a visa as his perfectly valid one had been issued in Paris. The local officials hadn't seen this type before and didn't like the look of it. He had to recross that damn ferry and face the dirt road all 500 miles back to the visa department in the capital. It is always wise to pretend to retreat, wait out of sight until another set of officers come on duty and then try again.

After our little altercation at immigration, Ken and I caught up with the others at the next river crossing. In this case the ferries were flat rafts. All Congolese ferries are supposed to be free but the locals don't think that way and refuse to take anyone across unless they are given a 'cadeaux'. Our convoy sat for half an hour in stalemate, us refusing to give money and they refusing to paddle. Of course they won. After a whip round to collect some CAR money, which was all we had, we were transported to the other side of the river. Ken and I camped with the Combi-vans and truck while the two Landrovers went on ahead.

The following day the road deteriorated; there were big mud puddles, many of which I fell into and had to be dragged out. Ken came off frequently and damaged his gear-change lever, also the spline had stripped and we had to pack it with tape. The next few days, battling through, trying to avoid mud holes and getting bitten was a nightmare. The jungle on either side of the road included huge bamboos with the shoots 6" in diameter and up to 40' tall. In some respects, although it was hard going, the bikes had it easier than the other vehicles because we could sometimes pick a way around and, when we did get stuck, it took far less time to pull us out. Everyone was helping one another. Even when we reached firmer ground things were difficult.

"I can't stand it any longer, I wasn't meant for this kind of life –

61

look at me – I'm filthy – bloody motorbike!!" For the hundredth time Ken threw his bike into the bush and swore he wouldn't get on it again. The poor thing wasn't built for the kind of punishment he and the roads were giving it and every time he fell off and something broke he lost his temper even more wildly. I'd try and remain calm while I attempted to repair the damage and talk him into a few more miles of being jolted up and down. We had terrible rows but I didn't dare leave him even when he said the rudest things because the thought of travelling alone frightened me to death and, even when the Landrovers and Combis said I could travel with them, I didn't feel it was right to let another soul travel completely alone especially as the Congo customs had pinched his tools and therefore he would have been sunk on his own. Despite these problems I found the going quite enjoyable in some ways; there were bright, iridescent butterflies all around us and a few interesting looking snakes wiggling across the road. Whenever we stopped I found a multitude of butterfly wings caught in the cooling fins of my engine.

We passed the Landrovers and caught up with the Combis, one of which had been stuck in a mud-hole for two hours. When the Landies arrived they towed the Combis through with the help of us all digging and pushing. I was now permanently covered with a layer of mud and never gave a thought to my appearance. Maybe these mud packs were good for my skin.

Fortunately the road improved tremendously into Buta. We were all in need of fuel and were surprised and delighted that there was a garage open with supplies as well as a bank but it did not accept travellers cheques. How frustrating! The elusive fuel was there but no money to buy it. I still had a few gallons left and hoped they would suffice until the next town.

Another incident occurred with the police. A pair of them stopped me and asked me to accompany them to the police station. I understood their French perfectly but had absolutely no intention of following them anywhere. I suspected it would result in some excuse for extracting money. I knew the routine now and smiled broadly at them.

"Yes, monsieurs, it is a lovely day. I am so pleased to meet you and you are both so handsome." Whatever they said I nodded and answered them nonsensically in English but made no attempt to move.

Into Africa – With a Smile

"I love your country and am riding on to Kissangani. Thank you for talking to me." More smiles. Eventually they looked at each other, shook their heads in exasperation and waived me on. Phew!

While in Buta Ken found someone with a welder at a Catholic Mission and they worked on his gear lever. The first attempt at fixing it failed and then the power went off and he had to wait for it to come on again before they could retry. No-one knew when this would be so I and the other vehicles carried on without him as he should be able to catch up with us by evening. I gave him a couple of gallons of fuel from my tank and hoped that we both had sufficient to make it to the next town.

The road was dry and firm for a change and I was getting up speed when I suddenly hit a patch of deep sand at about 40mph and fell off. As usual, I couldn't pick the bike up on my own but fortunately a group of locals appeared and with their help it was soon righted. However they demanded 'cadeaux'. I had nothing to give them and, mounting my steed, carried on only to fall off again about ten yards further up. As I had not come up with a 'cadeaux' before, my previous helpers just stood and watched but I wasn't worried as I knew our other vehicles were behind me and would soon catch up. This they did and I was remounted and on my way.

We crossed yet another ferry en route to Benalia and camped at a nearby Catholic Mission but there was still no sign of Ken. I was worried about him as his chain had been jumping off. Both the chain and sprocket were badly worn and with no tools he may be stranded somewhere. So the next morning I went back to wait by the ferry but, luckily he showed up just as the others were leaving. They had kindly given me enough fuel for the two of us to continue and had hurried on to reach Kisangani before the banks closed. While I was waiting for Ken to come over on the ferry the local children, with little brown, smiling faces were trying to outdo each other giving me different fruits and sweets and they all wanted to shake my hand. What a difference from the previous encounters!

It was only 70 miles to Kisangani and the road was good so we re-met the others at the P.O. and went to camp behind the Olympia Hotel where we were surprised to find the blue truck. They had been there a week as they had three cases of malaria on board and were waiting for them to recover.

Chapter Fourteen

Kisangani

It was wonderful to catch up with my friends on the blue truck, especially Jacky, and they were able to pass on information about where to get money changed on the black market at almost double the bank rate. They also recommended the coffee and peanut butter sandwiches in the bustling morning market which I found the best yet. It had a fascinating array of foodstuff which could not be found in Europe; bugs and insects like fried locusts were a tasty snack and, for something more substantial, a whole cooked monkey. These were a little disturbing to look at as they had a screaming expression on their faces as though they had been plunged live into a cooking pot. They looked like little human babies and I'm sure, in the middle of a sleepless night with a screaming infant, some mothers might have considered this option. The busy market was colourful with the local women in their bright cotton clothes and the array of exotic fruit and vegetables. We spent many happy hours there, amid the smell of exotic spices and, between mouthfuls of peanut butter sandwiches, Jacky filled me in about the events on the truck.

We had been on the road a couple of months now and, with the close proximity, different liaisons had formed as personalities had clashed or been drawn together. Apparently a few people had become disillusioned with Terry's leadership skills and the way they were always running out of petrol. A mutiny had almost occurred. However, Dennis, the driver had sorted it out and there was an uneasy peace though some of the group now cooked and camped separately. Jacky was happy to have my company for a while as we explored the derelict town together which still showed the scars of the troubles that had a occurred less than ten years before.

Zaire had been under Belgian rule as 'the Belgium Congo' up until the early 1960s and then there was internal trouble resulting in the Simba uprising. These were a breakaway terrorist group, very

INTO AFRICA – WITH A SMILE

violent. The conflict finally came to a head with the involvement of the Americans executing a daring airlift 'Operation Dragon Rouge' to rescue European hostages held in Kisangani then known as Stanleyville. Unfortunately, during the evacuation thirty hostages were killed by the Simba including an American medical missionary, Paul Carson. The event attracted worldwide television coverage and a book about it was written by David Reed, *111 days in Stanleyville*. A copy of this book was being passed around the truck passengers and I managed to read it. It certainly made us feel nervous especially as the European style buildings, once so grand, were now in ruins, evidence of a horrific conflict. The town had a fascinating but ghostly feel about it.

The Olympia Hotel, by which we were camped, had, in colonial times, been a high spot but now it was dead and drab. Many shops were closed for good and the ones that remained were very poorly stocked. There was one supermarket but that too had only a small selection of goods.

I met a young American who was doing a great trade on the black market changing money and also dealing in ivory. He was having to be very careful as the police were watching him. With a mop of tousled blonde hair, blue eyes and an open smile, Charlie had an engaging manner and was happy to show me around the ivory workshops where the locals carved some truly beautiful pieces from the elephant tusks. Though the finished product was attractive the story behind it isn't.

Trading in ivory had begun between African people way before the Europeans arrived, however, when they did there began a pattern of exploitation which was even witnessed by David Livingstone. Villagers were forcibly rounded up into camps and then ransomed through relatives who were forced out on elephant-trapping expeditions in order to be re-united with their family.

During the peak of the trade, during colonisation, around 800-1000 tons of ivory were sent to Europe alone. Ivory was desired by the Western world for piano keys, billiard balls, knife handles, inlay in furniture and ornamental carving. It was traded with northern America in exchange for calico which was popular in the Congo basin.

In the 1970s, Japan, relieved of trade restrictions, also started to buy up hard ivory which was popular for making name seals. Softer ivory was used for souvenirs and trinkets. At that time Japan constituted 40% of global trade and Europe and North America 40%. Hong Kong became the largest trade hub.

In the early 1900s shooting elephant and other tusk animals such as hippo, walrus and mammoth, was considered heroic and such slaughter continued in the sport of big game hunting as well as the quest for ivory. The elephant population had fallen from 26million in the 1800s to 1.5million in the 1970s. So Charlie's business was on the dark side although, at that time the ban on ivory trading had not been brought in.

A few of us went with him to visit the local zoo where there was an okapi, an animal which is native to the Ituri rainforest area in the Congo. This is very peculiar in appearance and looks like a cross between a zebra and a giraffe and is, in fact, a relative of the latter. It is about the size of a small horse with a dark-reddish back with striking horizontal white stripes on front and back legs. The neck is similar to a giraffe but shorter. Males have short, skin-covered horns and large ears to detect the sound of their predator, the leopard. In the late 1800s, the British governor of Uganda, Harry Johnston, discovered some pygmy inhabitants being abducted by a showman, for exhibition. He rescued them and in return the grateful pygmies helped him find out about this strange animal which had previously been noted as some kind of horse. He never actually saw an okapi live but found some skin and a skull which he took back to the UK where it was classified as belonging to the giraffe family and named *Okapi Johnstoni*. While at the zoo we also saw a baby elephant with which everyone fell in love.

In the evening Charlie and I went to a village on the river where the locals caught fish by building nets of basketwork which they placed across the rapids. It worked by a complicated vine-pulling system to trap the fish. Every evening they pulled the nets and sorted the catch. We helped them with this and were given a present of some small fish. The older men who have been doing this all their lives have marvellous muscled physiques.

Our little company spent nearly a week in Kisangani. The Welsh boys had given a young Frenchman, Loic, a lift earlier on the road and he was continuing with them. They left after a couple of days. I would have loved to stay longer but it was important to get out of the Congo before the big rains started as otherwise we would be stuck for about four months as the roads became impassable until the next dry season.

It was at this point that Ken decided that he was going to continue his journey south by boat on the river. The chain and sprocket on his bike were beyond repair and he couldn't get spares for it in Kisangani. Also, not being used to riding bikes, he was fed up with his continuous falls and being soaked through all the time. With no carnet to worry about he was going to sell the bike for what he could get for it. I must admit that I was relieved that I wouldn't have to worry about him all the time. We had managed our strained relationship tolerably well but he was a liability and becoming an expensive one. So we made our farewells and I continued in the company of one of the Landrovers. This was driven by a young couple by the names of Artie and Val. Artie was American and proud that his wife, Val, was Canadian as he said that was the next best thing to being British. He told me that in America he owned a Vincent motorcycle. How impressive!

The blue truck had continued on its somewhat troubled journey and the other Landrover and Combis had left to follow another route. Artie and Val and I set off together but, alas, the rains had already started, in fact we had to delay our departure a day after a violent storm during the night had turned the roads to such a slippery slime that it was even difficult walking to the market for breakfast. The next day it had dried out a little and we did about 170 miles to Bafaswende. The road was mainly fast going but with many hidden potholes, one of which caught me unawares and I ended up being stuck fast in the jungle at the side of the road waiting for Artie and Val to pull me out. That night and the next we stayed at a Catholic mission, camped outside.

At Mambas there was a tremendous storm during the night and my tent completely collapsed under the force of the water. The thunder and lightning were frightening and I thought the whole world was crashing down around me. I had to abandon the tent which was a flattened, sodden mess of canvas and poles, and run for shelter in a nearby cane hut.

In the morning the track was dreadful and I was sliding all over the place. I had to stop and take the front mudguard off as the mud was jamming between it and the wheel. I also had to slacken off the rear mudguard to lift it. Of course, with no guard on the front the mud was coming straight off the wheel onto me and I was covered from head to foot. With just my Barbour jacket and jeans and open face helmet I had no real protection. What a sight! It was all Artie and Val could do to keep a straight face when they looked at me. They had trouble getting the Landrover through the heavy mud and, at one stage, had to be towed out by a truck. The trucks became stuck too; it was not unusual to see six or seven in a row spending all day digging each other out and camped in a group overnight chatting and laughing. It was difficult to get past them, even if I could keep moving myself. I often had to enlist their help. In one place we were held up for a couple of hours behind a queue and were lucky it was only that long.

One day, while progressing in our slow fashion along the track, Artie and Val and I stopped to talk to a group of little people who had appeared from out of the deep jungle. They were Twa pygmies who lived in the area as hunters and gatherers, still free at that time. They were four to five feet in height, the men wearing cotton loin cloths and the women woven wrap round skirts and bare-breasted. The men had spears and little bows and arrows. They wanted to trade a weapon for my T- shirt even though it was dirty, tatty and full of holes. I would have loved to trade it for a bow and arrow but it was the only one I had so I couldn't. They were all smiles, however, and their little round wrinkly faces reminded me of elves. Val rummaged through the Landover and traded one of Artie's T-shirts so the little tribe were happy and disappeared into the bush waving us goodbye and we continued on our weary way. I for one was thrilled with the encounter and they had let me take photos.

Eventually we reached a hotel at Mt Hoyo and there I celebrated by having a much needed wash but had to replace the dirty clothing as I had nothing else. After the previous hard travelling we decided to stay for a day's rest at the hotel. Chatting to Artie and Val I discovered that, though they were married, Val had kept her maiden name. I thought this very avant guarde!

There were some caves locally so we joined some other tourists and drove out to see them. Artie and Val gave me a lift in their Landy and I was surprised to find it even more uncomfortable than my bike; I was bumped up and down, occasionally hitting my head on the roof. All in all I definitely preferred riding the bike than being in a Landy, despite my penchant for falling off. I replaced the front mudguard in the hope that the road would be better, with less mud to worry about. It transpired that there were still many deep potholes filled with water, sometimes coming over the cylinders but the trusty BM kept going.

CHAPTER FIFTEEN

Through Zaire to Rwanda

We finally arrived at the American mission at Hoysha where we stayed two nights trying once more to get clean and recover enough to face the next section of road which, we kept promising ourselves, would be better. In comparison it was, but was still pretty rough. Unable to get petrol at Beni, as they didn't want to sell us any, we continued to Butembo where we had a contact to look up. His name was Chris and he was a Peace Corp worker, a friend of some others we had met in Kisangani. Luckily he was in and even had a fridge stacked with cold beer which he offered us. What a treat!

Chris was able to sell us some petrol that he'd been able to obtain on the black market. Although it was three times the pump price, at a pound a gallon, we were only too pleased to buy it as I was completely out.

Staying overnight I was able to use Chris's garage in which to do a very overdue oil change. I had just bought a gallon for this. The old BMs used straight 30 grade oil all through the system: the engine, gear box and final drive. At this time this oil was available anywhere so it was a very practical idea. Using an impromptu funnel, made from a cardboard box found nearby, I managed to spill oil everywhere in the process and spent more time cleaning up after myself than doing the oil change!

The next few days the travelling was over very different terrain as we were approaching a mountainous region. It was rocky, high ground, still raining and, because of the altitude, much cooler. With no waterproofs, I was soaked and therefore felt even colder. I also felt miserable. I couldn't do more than 15mph and, even at that speed, had to stop and wait for the Landy. Being bumped up and down continuously my arms were aching with the strain of keeping the bike going straight. Altogether I felt sorry for myself and had a little cry while I was riding along.

Just as we were approaching Rwindi National Park I had a puncture in the rear tyre. Horrors! I had never mended a puncture myself as there had always been someone else around to help me. Thus I waited for Artie and Val to catch up hoping Artie would be my knight in shining armour as he had told me he owned a motorbike. However his response was:

"Hell no, Linda, I always take my punctures down to the mechanic to fix".

So, shamed into it, I began. Luckily the old BMWs made rear wheel removal easy as the rear mudguard stay doubles as a stand when loosened and lowered. Artie did help me get the wheel out. Then I attacked it with my one tyre lever, a big screwdriver and a spoon. I had a couple of spare tubes; the sensible BMW had 18" both front and rear, so I put one in and used Artie' stirrup pump to blow it up. Wonders, it worked! It had only taken me about 20 minutes whoa to go. That made my day, I couldn't believe how clever I was.

By this time it was late in the day, in fact early evening, the time when animals start moving around. As we were now in a national park it wasn't a good idea to dally. Indeed, the three-foot-high grass growing around us on the valley floor could hide any number of lions, hippos and other dangerous animals. We had to push on to get down a long escarpment with a rocky surface, hairpin bends and plenty of evidence of trucks going over the side. The plain stretched out below into the darkening distance and a few miles along we came to the Rwindi Game Lodge. It was very up-market and pricey but necessary to book in as camping in the game park is not permitted. In fact, with animals like elephants, lions and hippos around, open vehicles (such as motorbikes) weren't allowed either, but there was no other road. Anyway, it was a rare treat to stay in such surroundings and I must admit I do enjoy entering flash hotels looking scruffy and getting dirty looks from the staff, then emerging from my room later looking clean and respectable. In this case, though, without a change of clothes all I could do was brush off the mud. Not quite the same effect.

The meal of local game, probably Johnson's Gazelle, was delicious and the next morning we tucked into a breakfast of fruit, cereal, bacon and eggs. Also we were able to get petrol at pump prices which was a great relief. We filled all our tanks and jerry cans.

While stopping to look at some hot springs another vehicle drew up and we were informed that, as we thought in a game park, no-one was supposed to dismount from their vehicles. We learnt later that the Welsh boys, in their Combi, were fined for doing so. Just as well a ranger didn't spot me.

The road up into the mountains, the last seemingly never ending stretch to Goma was very bad with large rocks, just like an ISDT trials course. I thought the tyres would be cut to ribbons especially as they were getting low on tread but they survived.

The town of Goma is beautiful as it is on Lake Kivu, one of the deepest lakes in the world. However, this lake is potentially a disaster waiting to happen. It has huge quantities of dissolved gas held at pressure in its depths. If one of the nearby volcanoes erupts sending molten lava into the lake the resulting explosion would destroy hundreds of kilometres of land around with people, plants and animals. Populations of Uganda, Rwanda and Zaire would all be affected. Thank goodness it didn't happen while we were visiting.

When we drove into the town I was amazed to find the blue truck and the Welsh Combi still there. They had been waiting five days for petrol to arrive in town and were in the process of having a full scale battle to get it at the pumps. I stopped for a chat with Jacky and the others while Artie and Val carried on to the border, then I followed, stopping at the office in company with the truck.

Everyone had been smartening themselves up as there had been tales of long-haired, scruffy people being turned back at this border. It seemed that the officials took great delight in barring people from entering their country – for the craziest of reasons. The week after we went through we heard that they hadn't allowed any hitch-hikers and another time they decided to give Germans a hard time. As with other borders it is better to wait for a change of guard. With everyone on the truck in their best clothes and all the boys having had a hair-cut the night before, they were all set but I didn't fancy my chances in my mud-soaked jeans, hair completely matted and the bike a big lump of mud. However, it must have been one of their better-humoured days, as we were all allowed through, though only after being thoroughly searched.

Although it wasn't allowed, we camped that night by the side of the road halfway up a hill. After a brief rainstorm, during which I became soaked again, the sky cleared and was filled with the red glow from the nearby crater of Mt Nyiragongo, to the lip of which some

INTO AFRICA – WITH A SMILE

of the others had climbed while waiting for petrol in Goma.

We were now in the area of the great gorilla population and nearby was the Karisote Research Centre where Dian Fossey, with zoology students, was studying these animals. Dian was an amazing woman. Born in 1932 in America she was interested in animals from a young age and became an expert equestrienne. While working and training as an occupational therapist, she attended a lecture by the archaeologist, Lois Leaky and became interested in Africa. Borrowing $8,000, she took out her life's savings and went on a tour of Africa which included the Congo where George Schaeffer, the zoologist, had studied gorillas. Meeting up again with Leaky and his wife Mary, she was introduced to Walter Baumagant who was a hotel owner and advocate of gorilla conservation in Uganda which he saw as a tourist drawcard. It was while staying with him in Uganda that Dian first saw gorillas.

Leaky encouraged Dian to undertake a study of the gorillas in the same manner of Jane Goodall with chimpanzees in Tanzania so, after studying Swahili and primatology, Dian acquired an old Landrover and began field study in 1967 in the Congo. Initially she lived in a tent amongst the gorilla population and her ease of communicating with them she attributed to her earlier experiences working with autistic children. She became known as '*Nyirmachabelli*, the woman who lives alone on the mountain.'

Because of political troubles both in the Congo and Uganda she decided to move and continue her studies on the Rwandan side of the Virunga Mountains and founded the Karisote Research station near Ruthengani. We could not visit this centre as Dian did not allow tourists believing that the animals were susceptible to human diseases such as flu. She only had a few students staying there. Dian encountered terrible trouble with poachers who would kidnap young gorillas to sell to zoos. The adults would fight to the death to protect their young and many animals, including those she had become close to, were killed. The poachers would take all the adult body parts to be used as ashtrays or magic charms. In later years, after we were in the area, World Wildlife Fund and other supposed preservation societies tried to wrest control of the Research Centre from Dian, saying that she was unstable. They wanted to control it for purposes of tourism.

In 1983 Dian wrote a book called *Gorillas in the Mist* which was made into a popular film but in 1985 she was found murdered in her

73

cabin. Her death was attributed to the poachers she was trying to eliminate although no one person was ever accused.

It was a shame we missed seeing these remarkable animals but I respected Dian's concern for them.

CHAPTER SIXTEEN

A Long Way Round, Wisely

A few years earlier a more simple overland route to East Africa would have been to travel due east from Kisangani in Zaire to the border of Uganda then drive over the nicely-sealed roads to the capital, Kampala, continuing on the bitumen through into Kenya and down the Rift Valley to Nairobi. However, in 1974 this route was impossible because of the regime of terror that was currently imposed by the dictator, Idi Amin, who had seized power in 1971.

Uganda was colonised by the British in 1800 and, over the years they had trained Africans to serve in the British Colonial Army. During the 1950s Idi Amin, a very fit, 6'4" young athlete had been recruited and was a popular rugby player. He was also a light/heavyweight boxing champion and a strong swimmer. He fought in Kenya against the Somali rebels in the Mau Mau crisis and in 1959 was made a warrant officer, the highest rank for a black African in the British Colonial Army at that time. He was soon promoted to lieutenant and then in 1963, a year after Uganda's independence, he became a major.

The Prime Minister at that time was Obote and he made Amin commander of the army in 1965. This proved to be Obote's downfall. Initially he and Amin were great pals but Amin eventually fell out with him and in 1971 arranged a military coup and gained power as a dictator, putting Uganda under military rule. Obote took refuge in Tanzania and the following year attempted a coup to regain power but it failed. Amin went on to reign with terror. There was an atmosphere of violence throughout the whole country and people were killed indiscriminately. Bodies were often dumped in the River Nile. Asians were expelled and many with British passports escaped to the UK.

In 1972 Amin broke diplomatic ties with the UK and nationalised British-owned businesses. In 1973 the USA embassy in Kampala closed. In 1974 while I was on the road, it was impossible to enter Uganda (even if I'd wanted to!). Amin's exploits had received much publicity. He was depicted in many Western newspapers as a buffoon, a figure of fun, dressed in his army uniform with his medals both real and self-presented but the horrific reality of his regime was only too obvious and the people who could, were leaving Uganda in droves. The Africans took refuge in nearby Tanzania, Rwanda and Kenya and others, with British passports travelled to Britain or Europe. Amin began discussing plans for war against Israel using paratroopers, bombers and a suicide squadron. The incident, when a plane taken over by Palestinian terrorists landed at Entebbe airport in 1976, aided by Amin, was made into the films: *Victory at Entebbe* and *Raid on Entebbe*.

Finally, by 1978 support for Amin had shrunk because the economy and infrastructure had collapsed from years of abuse and neglect. He was in conflict with both the neighbouring countries of Kenya and Tanzania and several of his ministers had fled into exile or defected. Even his vice-president mutinied. In 1979 Tanzania mobilised an army with Ugandan exiles and Amin was forced to flee by helicopter. He escaped to Libya and ultimately to Saudi Arabia where the royal family gave him refuge in exile and an allowance on condition he kept out of politics. In 1980 Brian Barron, a foremost BBC African correspondent, located Amin in Saudi Arabia and interviewed him. In the interview Amin held the view that Uganda needed him and he never expressed any remorse for the nature of his regime during which about 300,000 Ugandans had lost their lives.

In 1989 he attempted to return to Uganda reaching Kinshasa in Zaire but the president, Mobuto, forced him to return to Saudi Arabia. In 2003 one of Amin's wives, Madina reported that he was in a coma and near death from kidney failure in a hospital in Jeddah. She pleaded with the president of Uganda, Museueni, to allow him to return to Uganda to spend his remaining days there but Musueuni said that Amin would have "to answer for his sins the moment he was brought back". Wisely Amin's family decided to keep him where he was and on 16th August 2003 gave permission to disconnect his life support. His remains lie in a cemetery in Jeddah in a simple grave and the funeral was conducted without fanfare.

Idi Amin was a large man both physically and figuratively. He was reputed to be either very charismatic or insanely violent and it is speculated that his extreme behaviour could have been bi-polar or the result of syphilis. For a time a British man, Bob Atkins was a close confidant and his role was implied in the film *Last King of Scotland*. A book about Amin's regime is *A State of Blood* written by his former health minister, Kyemba, who defected.

Obviously, given Idi Amin's exploits and reputation in 1974, I was only too happy to take the long way round through Rwanda.

LINDA BOOTHERSTONE

RWANDA TO KENYA

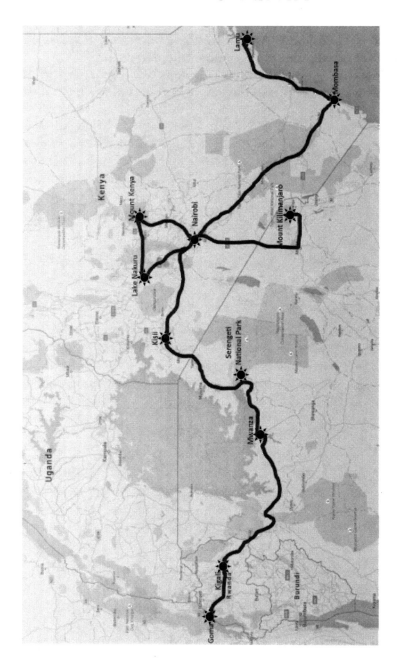

CHAPTER SEVENTEEN

Rwanda

Rwanda had its own problems. It had previously been ruled by the Belgians who had favoured the Hutu tribe. Since its independence in 1962 there was already conflict between them and the other major tribe, the Tutsis although they were integrated in many areas. The time I was there, in 1974, the improved economy, prosperity and decreased competition for work meant that there was less violence between them. However, increasing population meant competition for land and this was evident to me. The countryside was intensely cultivated and I saw many people working the fields alongside the road. There were no open areas where we could camp as in previous countries.

Leaving the others after camping with them on the open road the night before, I continued alone in the morning, the bike speed being far greater than that of the truck over the hilly terrain. The surface was bumpy but a great improvement over that of Zaire and the scenery very pretty with lush green vegetation neatly organised.

I had arranged to meet up with the truck in Kigali, the capital, and was happily travelling when, about 50 miles short of that town, I saw a BMW motorbike parked by the side of the road. Upon investigation I found it belonged to a Swiss pastor who was working among the locals. During the course of the conversation he explained to me one of the customs of the people as I had noted that the majority of the field workers were women.

"Why is this?" I asked. "Where are the men?"

"Ah, but the men are in the drinking houses, of course, debating with their friends," he explained. Discussing politics was considered very important and if the men did not attend these sessions they were looked down upon by the others. Tending the banana plantations was considered women's work. He said land was scarce; the population had swelled not just with normal increase but with refugees from troubled Uganda.

Reaching Kigali, a sprawling city set on several ridge-tops around a valley, I needed to find a bank to change money. When I did so I was told it was closed and it 'might' open after lunch so I sat on the pavement opposite the market and was soon surrounded by young men. Some of the boys were selling peanuts but I indicated by pulling out my empty pockets and pointing at the closed bank that I didn't have any money. One young man stepped forward and bought some nuts for me which I tipped into my crash helmet. He was so pleased to see me eat them he bought some pieces of meat on a skewer and then rushed off and bought me some tonic water in a beer bottle. He was obviously showing off in front of the others but I didn't mind as I was rather peckish. The group of boys just stood around watching me eat until the owner of the shop outside which I was standing became annoyed at the blockage they were causing and chased them away.

Finally the bank opened, I was able to get some money and I went to fill up with petrol. This was a lengthy process as it appeared I had to apply to the council for a permit. This achieved, I sat in a queue at the pumps all afternoon and I saw Artie and Val also queuing. They had been there all morning but didn't know about the permit and when they finally made it to the pump were turned back to get one. So, they had to start all over again the next day – this time with the all-important permit. Even then there was a fight and Val ended up throwing a bucket of water over one of the pump attendants.

I went to the Hotel Relais to meet the others on the truck, as arranged, but there was no sign of them. I was just wondering where to stay when I saw Artie and Val and went to the Catholic mission to camp with them. My pannier brackets had come adrift on the Congo roads and I needed to find somewhere to get them welded. Fortunately, behind the mission was a workshop run by a Spaniard, Monsieur Albert. Lo and behold, when I walked into the workshop there was a nest of BMWs like mine! They belonged to the missionaries and he did all the work on them and was quite an enthusiast himself. He welded my brackets and patched my inner tube for a small fee so, thus prepared, I set off to buy some food before starting my journey out of Rwanda.

I was just leaving the market when I saw Terry and Jacky from the truck walking down the road carrying a huge leaf spring. It was from the truck and had broken about 30 miles down the road out of town so they had hitched a ride in to get it mended. I took them to Monsieur Albert where the repair was soon made and it was arranged that Terry and Jacky stay in town while I took the spring back to the truck on the bike. What a sight! This huge spring lashed across my rear seat. I felt like one of those balancing toys and it made the steering a tad tricky but I duly arrived at the stricken vehicle, Dennis fitted the leaf spring with the help of the others and they followed me back into town. I spent another two nights in town with the gang. The Welsh boys and a French hitchhiker, Loic, they had picked up in Zaire were also there and we danced the night away. The young Rwandans were fabulous dancers, of course and we really let our hair down.

Travelling alone once more, I reached the Rwandan border to exit the country by mid-morning. There was no sign of the customs official and, had it not been for the army personnel hanging around and the fact that I needed the carnet stamped, I was tempted to ride straight through. After about twenty minutes, someone official arrived looking very bad-tempered and I started filling in the necessary form. However, he suddenly asked me for 200 francs

"What for?" I demanded.

"You must pay to cross the bridge." This was the bridge that spanned the river adjacent to the customs post and which led to the Tanzanian border.

I was quite taken aback and accused him of wanting the money for his own pocket. He became very upset, denied the accusation and showed me a receipt book. I didn't have any Rwandan money and was loathe to give him any American dollars as they were worth so much more, so I told him I would have to wait for my friends to arrive to pay for me. He shrugged his shoulders and marched off up the road leaving me alone to wait. I wandered over the bridge to take photos of a waterfall and narrowly escaped arrest by the army as, of course, it is illegal to take any photos on a border area. Silly me. I played innocent, unable to understand French again and they left me alone.

Eventually Artie and Val arrived at the post carrying another corpulent, scruffy-looking customs official to whom they had been forced to give a lift. He reeked of drink, had smoked like a chimney and had used their wing-mirror to preen himself all the way. He had also spent the whole time making rude remarks about Val and they were altogether cheesed off with him.

As he alighted from the Landrover and we made for the customs shed he pompously announced, "Closed for lunch, back in an hour."

He picked up my motorcycle gear which I had left in the shed and threw it out. As it started raining again and there was no other shelter, I just had to stand around and get wet. Finally he re-opened, made more rude remarks but stamped our carnets and we were free to go. Then we had to wait another half-hour at the Tanzanian border up the road as the official there was having a bath in the river. However, he spoke English and was very pleasant. After one more stop for immigration we completed 40 miles to the mission at Rulenge just as darkness descended.

CHAPTER EIGHTEEN

Tanzania

As a treat we stayed overnight in a room in the mission and took breakfast together before investigating how we could get to Mwanza on Lake Victoria. This huge stretch of water is one of Africa's Great Lakes and was named after Queen Victoria by the explorer John Hanning Speke while on an expedition with Richard Frances Burton to find the source of the Nile. He was the first European to document it, although the initial records of this lake came from Arab traders in search of gold, silver and slaves. The surface area is 26,600 sq. miles and it is Africa's largest lake and the largest tropical lake in the world. Its shores lie on Kenya, Uganda and Tanzania but its only outflow, which is in Uganda, is the River Nile.

Speke and Burton's claim that Lake Victoria is the source of the Nile was later challenged by Henry Moreton Stanley (yes, the one that found Dr Livingston) when he noticed nearby snow-capped mountains that were normally hidden by cloud. Called by locals, the Mountains of the Moon, they are the Rwenzori Mountains and the run-off from their ice melt into the lake is some of the source of the Nile waters.

Unfortunately Lake Victoria's initial natural biological wealth has been destroyed by the introduction of Nile perch and water hyacinth. The latter thrives on the untreated sewerage that enters the lake and chokes it, cutting off the oxygen needed for the fish. More than 30 million people depend on its resources and now, in later years, steps are being taken to overcome these problems.

As we approached it we were unaware of this information, just knew that we could catch a ferry to take us across an inlet and into Mwanza.

There had been a great deal of rain, the road had been washed away in places and then, to my utter horror, we came upon a river, about 30 yards wide, running across the track. It was fast flowing and

I had no idea how deep it was. There was no way round and I just sat and gaped at it. Artie and Val drew up behind me and I turned around and grimaced. There was only one thing to do and that was to ride through it so I gritted my teeth and began. The water came right over the engine, up to the seat and, as I rode forward, a wave came over my head so I couldn't see where I was going. Spluttering and gasping, I kept moving, feeling the tyres sliding on the loose gravel beneath the water. I was slipping the clutch, with throttle open to keep the revs up hoping to stop the water entering the exhaust. Miraculously I reached the other bank and came to a halt, keeping the spluttering engine revving while Artie and Val came through in the Landy. They said that they couldn't see me for the spray and Artie again expressed his amazement at the fortitude of my bike.

We continued over windy roads and then hit some bad sand and corrugations as we came out of the mountains and onto the plains. Suddenly there was a crash and my panniers collapsed. I stopped, took out my tool-kit and removed them totally, putting them on the roof of the Landy. The bike looked naked without them.

We arrived at the ferry to cross Lake Victoria just as the sun was setting and were treated to an array of beautiful colours, all shades of red and orange, reflected in the water as we slowly crossed the lake. We were three very tired people who eventually drove into Mwanza, a huge town to our eyes. It even had a railway. First things first, desperate for a beer, we found the Mwanza Hotel and were amazed at the low cost of food and drink compared to the previous countries. The Tusker beer was very much to our liking and, after our well-earned drinks, we camped next to the lake.

The following morning I found an engineering firm to weld my pannier brackets. It was also a Honda agent, although all the bikes were of small capacity. The manager insisted that I let all the boys do the work while I joined him for a drink in his office. He produced a bottle of whisky and I think he had the idea of getting me tipsy as he had that look in his eye. However, an old hand at whisky drinking, I showed no sign of inebriation and he kindly gave me the rest of the bottle to take with me on my journey.

The blue truck arrived the next day and they had a great deal of work to do on it so I started out on my next stage alone. The truck, Artie and Val and the Combi with the Welsh boys were going to take the road through the Serengeti National Park into Kenya but this I could not do.

One of the most well-known national parks in Africa, Serengeti, spans the border of northern Tanzania and south-west Kenya and hosts the largest terrestrial mammal migration in the world. There are seventy large mammal species including blue wildebeest, gazelles, zebra and buffalo and it is renowned for its large lion population. The Serengeti NP was the inspiration for the film *The Lion King*. The park has a varied landscape of swamps, kopjies (hillocks), grassland and woodland and hosts 500 bird species. The terrain ranges from short treeless plains to large areas of Acacia trees and is home to the Masai tribe. In fact the word Serengeti means 'endless plains' in their language. Of course, open vehicles such as a motorcycle were definitely not allowed to pass through the park so I had to circumnavigate the area. It wouldn't do to be swallowed by a lion!

However, I camped at the entrance to the park that night and was not a little worried by the strange noises very close to the tent. The zip on the entrance had broken and I couldn't close it. This made me feel very vulnerable – though a tent would be scarce protection against any marauding animal. I heard a lot of crashing branches and other terrifying noises during the night but I just closed my eyes and crossed my fingers. The next morning the park warden told me there had been hippos around so I was lucky – perhaps the smell of my unwashed body didn't appeal.

I did a quick check on the bike and then continued to the Kenyan border. While crossing another ferry, I met a Danish boy who was the equivalent of the American Peace Corp workers. At a tea shop a little later, I met a young American who was living in the area doing odd jobs and it was while chatting to him that I tasted my first samosa, an indication of the Indian influence in this part of the world. Finally I checked out through the Tanzanian border and, at the Kenyan border, came on to the wonderful, marvellous tarmac roads. I almost kissed the border guards with relief. They were very friendly, English-speaking and suggested that I went to the hostel in Kissi, a popular watering hole. This I did as I had good reason to celebrate – three months of dirt-road riding behind me, it was now all bitumen to South Africa. Not only that but now, for the first time since leaving Britain, I was driving on the left-hand side of the road. Not that it particularly worried me, riding on the right, but this made me feel more at home. At the hotel I was bought drinks all night by a charming, black Kenyan who had spent some time in England.

The next day I made a clear run through on the lovely, smooth surface to Kericho and then on to the Capital, Nairobi. My one worry was that the tank started leaking petrol around the fixings but I knew that I could mend this when I arrived at City Park, the Overlander paradise.

In Central Africa

In Kano, Nigeria

French Foreign Legion

In Zaire with Ken

Congo mud

Into Africa – With a Smile

Planning a route through the mud

Mike by a ferry crossing

Congo pygmies

Elephant tusk for ivory trade

Dismantling leaf-spring from the truck

Jacky with locals

Truck and passengers

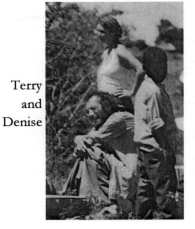

Terry and Denise

Motor Cycling UK, 1974

Dian Fossey with Gorillas

Serengeti National Park

Mount Kenya

CHAPTER NINETEEN

Nairobi

City Park, Nairobi. A home away from home, it was just a small piece of land within the city centre, set aside for travellers. Although it had only one toilet and shower, there was a terrific atmosphere of comradeship generated by stories of shared hardships.

As I rode in I received a warm welcome from the many people I had met along the route. All the Overlanders congregated there: those coming up from South Africa, those heading south and some who had come across from India and heading either way. What a variety of vehicles and people and tall stories, very often true, that people related about their journeys, all accompanied by music from guitars mingling with the dagga smoke that hung above the tents. The latter were pitched in a square formed by road-worn vehicles parked around the fence of the enclosure. I brought out my lagerphone and joined in with many singsongs. I also collected more bottle tops for it from the bottles of 'Lion' and 'Kilimanjaro' beer.

For many people this place had been home for several months as they waited for visas, the weather to change or to sell, buy or repair vehicles. All through the night there was usually someone working by torchlight and plenty of people to lend a hand changing tyres, springs or even engines.

My bike needed attention; I had to fix the petrol leak around the fixing. A bolt had bent and its head had penetrated the inner skin of the container. I had been given some especially strong epoxy by Don from Tridon Spares (my sponsor!). He told me to be very careful with it as it had two parts that have to be mixed precisely, as it heated up when creating the chemical reaction needed. I put a portion of each part on a piece of cardboard and then mixed them together in preparation for applying to the emptied tank. Then, fortunately as it transpired, I was diverted into conversation with some friends. When I returned my attention to the job in hand the cardboard was smouldering and just about to burst into flames. My mixing quantities were incorrect and if I had applied it to the tank there might even have been an explosion! Lucky me. I paid more attention to the remix and the tank was duly mended.

Nairobi was a beautiful city with all mod cons and, at that time, had a free and easy atmosphere with many tourist attractions. I visited a reptile park and was amazed at how many different kinds of snakes there were and how easily they were camouflaged in the bush. There were about 200 species of snakes and reptiles including tortoises, turtles and crocodiles. Some of the venomous snakes were housed in open pits. These were looked down upon from a circular, raised viewing platform, a safe distance away, and it took some time to distinguish the snakes from the branches of the tree in which they were resting. No wonder I hadn't spotted any along the road.

I had accumulated a number of cassettes of exposed 35mm slide film with unique pictures of me being stuck in the mud in Zaire, pygmies, elephants and my friends on the road and I was keen to post these off from the main Post Office to the film processing labs in the UK with my mother's address as the recipient. I was also keen to collect any mail waiting for me at Poste Restante. Of course, we all had to have a beer in the New Stanley Hotel, famous for its use by upper-crust whites in the colonial times.

I had plenty of other things to do while camped in City Park. My jeans, often beaten clean on river rocks by local women laundresses along the way, were ripped almost beyond repair but I persisted with them. With scraps of material discarded by others from their worn clothing, I sewed on many patches in varying colours. Now, about a stone lighter in weight, I also treated myself to a size-ten pair of red jeans from the local market. I felt very trim and sexy. While in the fashion mood, another woman and I spent a whole day plaiting each other's hair in the African style that was to become popular with young Westerners in years to come. I also patched my boots and tent and tracked down a long zip to fix the entrance.

During the first week I was in the park Artie and Val arrived but there was no sign of the blue truck or the Welsh combi. I was looking forward to their arrival as we could all celebrate and relax together now the worst part of the journey was over.

Food was wonderfully cheap in Kenya – great big steaks for 4/- a lb and I immediately started putting on weight. It wasn't long before

INTO AFRICA – WITH A SMILE

I had to find a way to let my new jeans out and reverted to wearing my old patched ones. Unfortunately though, my money was nearly all gone and I had to phone home for my last emergency pounds to be sent out.

CHAPTER TWENTY

<u>Mt Kenya</u>

While waiting for funds, I joined two Austrians, two French Canadians and a Swiss guy in their Combis and we went off to climb Mt Kenya. We drove to the entrance of the NP where we paid our money and hired a guide who carried our sleeping bags and some food in a large rucksack which dwarfed his slight frame. We then continued by vehicle up the initial slopes of the mountain to park and start the trek.

Mt Kenya is situated on the Equator and its highest point is just over 17,000ft. It is the second highest mountain in Africa, Kilimanjaro being the first. We didn't actually begin to climb until about 10,000ft and really should have started at a lesser height to acclimatise. By the time we stopped in a hut the first night the temperature was markedly lower and in the morning there was frost on the ground. None of us had proper mountaineering gear and I was walking in my motorcycle boots and jacket. The altitude meant that we were quickly short of breath and I felt that we were probably the slowest climbers our guide had ever taken up. As we staggered ever higher, one of the boys, Andre, the French Canadian, was physically sick and I was the only one, apart from the guide who didn't have a headache.

"Linda, you are my 'ero," Andre kept telling me as he struggled to keep up.

High above the Rift Valley the scenery, looking down on the wooded mountainside and the grassy plains, was spectacular and, as we reached the snowfields, even more so though the ground became all the more treacherous for climbing.

After three days of climbing we reached the last hut. We slept early and then rose before dawn to climb for an hour in the dark, over icy terrain, to the summit. On arrival we watched the sunrise as it gradually illuminated our lookout point and the high peaks above us, only accessible with full climbing equipment. The view was breathtaking; the surrounding clouds coloured all shades of pink and the morning mist clearing to give us sweeping views of the Kenyan landscape far below to the other world where the animals were roaming in amongst the acacia trees and buffalo grass. The descent didn't take so long although my legs became very wobbly with the constant pressure on my knees and I kept falling flat on my face.

While painfully negotiating the lower slopes, I was interested to see what looked like large guinea pigs, hopping among the rocky outcrops. They were brown in colour with short hair, small ears and a stump of a tail. They had padded feet and made a kind of grunting sound. These are rock hyrax, sometimes called rock rabbits or dassies. In fact the true dassie is smaller and more like a rat and not found on Mt Kenya. The rock hyrax has long, black whiskers and tusk-like upper incisors and is, in fact, the closest living relative to an elephant! They were fun to watch, their little noses quivering as they searched for food of grasses and bugs on the rough ground.

We were glad to reach the plains and spend a couple of days relaxing, camped by a farmhouse near the foot of the mountain. This was owned by a little, old Irish lady in her 70s, Mrs Keneally. Her large house, surrounded by a veranda, was just like that of Karen Blixen in *Out of Africa*. (In fact the house used for the film is a place called Ngong Dairy situated in Karen, a suburb of Nairobi). Inside this one, in a colonial-style room with many Irish mementoes, she had a piano. Andre, recovered from his mountain sickness, proved to be an accomplished pianist, leading us in a good old honky-tonk style sing-along in French and English. On the way back to Nairobi we visited Lake Nakuru and were delighted by the sight of flocks of pink flamingos. More worrying were the warning signs 'Beware of Hippos'.

Arriving back at City Park everyone welcomed us 'home' and my bike had been well guarded in my absence. However, I was sad to hear that I had missed the blue truck as they were behind with their schedule, having been stuck for days in the flooded Serengeti with the Combi. They had left for their destination of Johannesburg in South Africa. John and Mike, the Welsh boys, were still there but shortly heading for Malindi on the coast.

I was enjoying the company at City Park but, much as I would have loved to stay longer, my money had arrived and it wouldn't last forever so I had to make a move further south. There was a notice board in the park with adverts and recommendations for accommodation elsewhere and I jotted down a few addresses, took many fond farewells of my travelling friends and departed, heading toward the coast.

While I was in Kenya I was aware that Jomo Kenyatta was president, also that it had previously been a British colony and that there had been 'a native rebellion' by the Mau Mau in the 1950s. However, I was not fully conversant with the history of the country or Jomo's involvement until later.

Britain began colonising Kenya in the 19th century and, dispossessing the locals, soon set it up with tea, sisal and coffee plantations. Nairobi was founded in 1899 as a railway link between Uganda and the port of Mombasa to assist in exports of these goods. At 1,795m above sea level its altitude made it a comfortable temperature and, although the population had problems with malaria, the town soon became a centre of administration and tourism, especially big-game hunting.

Some of the less pleasant administration techniques of the British Colonial rule led to native dissatisfaction and, shortly after WW2, there was an uprising by Kikuyu rebels who became known as the Mau Mau. They targeted the British and those who worked for them and used terrorist techniques; its recruits were made to swear blood oaths to kill Europeans and their African collaborators. By the time the Mau Mau movement fizzled out, they had killed 13,500 Africans and 100 whites and many domestic animals. During this bloody reign of terror many white Kenyans left for good.

Jomo Kenyatta was born in Kenya in 1889 and, while a young man, became interested in politics. He was elected president of the Kenya African Union and lobbied on behalf of tribal land affairs both in Kenya and in London. He briefly studied Communism in London and Moscow and became viewed by the colonial masters as a trouble-maker, even receiving death threats from white settlers. The Mau Mau rebellion started in 1951 and in 1952 Jomo was arrested and charged with managing and being a member of the this group. The trial lasted five months and the evidence against him was shaky but the colonial powers wanted him out of the way and he was sentenced to seven years imprisonment with hard labour and was refused appeal. The Mau Mau rebellion was controlled in 1956 and in

Into Africa – With a Smile

1959 Jomo was released from prison but still detained in a remote area. In 1961 a public meeting of 25,000 in Nairobi demanded his release and a million signatures were on the petition.

When Kenya was granted independence in 1963, he became its first Prime Minister, hailed as the founding father of the Kenyan Nation, the 'Lion of Africa'. He supported reconciliation with the whites and consistently asked them not to leave. From 1966 onwards he made sure that he was kept in power by disposing of his opponents in sometimes shady fashion. As old age took its toll he became less involved in state affairs himself but his ministers, picked from his Kikuyu tribe, took over many of his duties, making sure that they were well paid. He was the only candidate for the 1974 elections and finally died in 1978, four years after my visit. I was aware, while in the country, that people were wondering if there would be an upheaval when Jomo died as there was opposition to the domination of the Kikuyu tribe in politics. However, I was unaware of any open unrest. While in Nairobi I found a copy of *Uhuru* by Robert Ruark which, though fiction, gave an insight into African affairs.

Nakuru National Park

Arty and Val in City Park, Nairobi

Near Serengeti

Nairobi

Nairobi

Hair in plaits at City Park, Nairobi

The new 'after-plaits' hair-do

Into Africa – With a Smile

Lion

Cheetahs

Tortoise at a reptile park

Above: Idi Amin
Right: Jomo Kenyatta

When the missionaries arrived, the Africans had the land and the missionaries had the Bible. They taught us how to pray with our eyes closed. When we opened them, they had the land and we had the Bible.

- Jomo Kenyatta

Kenya to Zambia

CHAPTER TWENTY-ONE

The Kenyan Coast

I took the route to the coast via Arusha, a small town in Tanzania which was famous for its meershaum pipes; hand-made and exported all over the world. Meershaum is magnesium silicate and the best quality of this substance is found in Turkey where it is of a creamy-white colour. However, it is also found in the Ambseli basin in Tanzania where it is brown, black or yellow. For some years it was mined there by the Tanganika Meershaum Corporation and a company called the Kilimanjaro Pipe Company set up its office in Arusha. It closed some years after I was there and the meershaum from this area was sent to make pipes in the Isle of Man. The material is used for pipes not only because it is soft and easy to carve to produce pipes with decorative shapes but meershaum absorbs the harmful elements of tar and nicotine. In doing so, the colour of the pipe changes, giving a darker brown, even wooden effect. It is sometimes used as a lining in briar pipes. I took a tour of the factory there, fascinated by the array of shapes and sizes of the carved pipes and bought one as a souvenir.

The next day I rode to Moshi in the foothills of Mt Kilimanjaro. I couldn't actually see it as the mountain was covered in cloud and was totally invisible. While riding through the heavily cultivated hinterland I was stopped by and invited to breakfast with an elderly Swiss man and his African wife. Over fruit, bread and delicious, aromatic coffee he told me that, when Tanganyika became independent in 1961 and, in 1963 came under the new government as Tanzania, he was very lucky to keep his little coffee plantation, as many of his European friends had been turned off their land that was appropriated for indigenous people. He was worried that in their later life, after having spent around 50 years in Africa and returning to Europe, they would have no land and maybe no relatives to help either.

Although very close to Kilimanjaro, I decided against climbing it

99

as my legs were still recovering from the climb up Mt Kenya and I thought that one mountain experience in a month was enough. Mt Kilimanjaro is higher than the latter, standing at 19,341 feet above sea level and has three volcanic cones. In 1848 a missionary, Johannes Rebman, working in Mombasa, reported its existence to the rest of the world and the first European ascent of the summit was made in 1889 by Hans Myer. Now it is a popular mountain to climb but is expensive to obtain permits.

As I was riding along a dirt track toward the coast I noticed, out of the corner of my eye, a large grey mass at the side of the road. Coming to a halt I looked back and, sure enough, standing there was a very large elephant with long, sharp tusks. I turned the bike and started back towards him to take a photo especially for my mother who loved elephants and collected many ornaments of them. As I approached, he trumpeted and stamped his feet so I opened the throttle and shot past, not bothering with the photo. But now I was facing the wrong direction and had to go back. The elephant obviously didn't like the sound of the bike and stood trumpeting and stamping at the roadside. He didn't look like moving. I waited a few moments then, gathering all my courage, opened the throttle and rode past as quickly as possible hoping that he wouldn't charge. He just made a good deal of noise as I passed but I just kept on going and decided that my Mum would have to do without any photographic evidence, at least not at that range. My cheap camera did not have a zoom.

Arriving at Mombasa, the fascinating port town, I tracked down an address I had seen on the board at City Park. This was about ten miles north of town and entailed slowly negotiating sandy tracks, through palm trees, to find a small house with a few outbuildings set up on a cliff above a beach. It looked out over the sea where lay a coral reef. A gentle breeze blew keeping the area cool.

Miki, a young Japanese man in his mid-twenties, was in charge. He was a botanist and was involved in a study of mangoes for the Agricultural Research Centre and had the house supplied with his work. To make a little extra money, he was inviting travellers to camp in the extensive gardens, among the fruit trees, for only six Kenyan shillings a night. There were always a few travellers coming and going, either camped in their own tents and vehicles or taking shelter in the outbuildings. A good deal of them were Japanese and it was interesting for me to be in their company as it was the first time I had had contact with this culture. They loved playing chess (I am hopeless at it) and they showed me origami and made delicious food. The local fisherman called daily to offer us their fresh catch of lobster, octopus and tropical fish and I learnt to eat fish raw with soy sauce and side dishes of rice and vegetables. According to the Japanese there, Japanese soy sauce is vastly superior to the Chinese version and Miki and his friends bought it in large quantities as no meal was complete without it.

Miki's house was named Pole Pole (pronounced polee, polee) which is Swahili for slowly and it was easy to see why – my life style slowed almost to standstill and, with all the tasty food, I gained even more weight. The coral reef provided hours of entertainment: snorkelling, looking for shells, crabs and other marine creatures but one had to be wary of cuts to the feet by the sharp coral and Mercurochrome was much in demand. One night there was a big storm and a large fishing vessel with refrigerated storage for deep sea fishing, came aground on the reef. The Japanese were over the moon. If they could get out to the ship across the coral, huge fish were being given out by the crew as the electricity to keep them fresh was not working and they had to be disposed of. I think the ship was finally classed as a wreck as it was abandoned and didn't get pulled off the reef all the time I was there.

Miki, a gregarious young man, took a shine to me and insisted in engaging me in conversation to improve his already-very-good English. He proudly informed me that he could say 'I love you' in ten different languages. He also loved to hear me sing and insisted that I teach him how to play my lagerphone. Having a good sense of rhythm, he mastered it surprisingly quickly and began demanding that everyone save their bottle-tops so that he could make his own.

He had been in Mombasa for a couple of years and had made some African friends. One night he took me to meet them and we had a meal together, squatting around the blanket on the floor upon which were a variety of dishes. One was fried beetroot which to me was far more palatable than the normal pickled sort. We used our hands to eat.

One embarrassing thing happened while I was staying there. Many people were coming and going and, in one of the outhouses, I found a small suitcase that no-one claimed. I opened it and found some women's clothing including a pretty embroidered blouse. As there were no other women in the camp and hadn't been for some time I appropriated the garment and wore it. Imagine my embarrassment when the owner shortly re- appeared and demanded its return!

To enlarge my much depleted wardrobe I bought a cotton sarong, a very popular garment in this area. Mine was turquoise with yellow edging. Instead of wearing it in the normal wrap around fashion. I folded it over length wise and cut a hole for my head adding my patched jeans when riding on the bike.

While at Pole Pole I heard about an interesting island about 200 miles further up the coast, Lamu, and decided that I would like to visit it so I set out, calling in at Twiga Lodge at Malindi on the way in the hope that I would find the Welsh boys there. Sure enough, there was their Combi but only one of the Welsh pair, John, was camped nearby and he told more about the duo and what had transpired since I last saw them.

The boys were from a small town near Swansea and had been friends for years before deciding that, at age 25 they should stop talking about overland travel and go out and do it. John worked in Barclays Bank and Mike was an electrician. They had prepared the Combi and left the UK around the same time as myself to traverse the whole of Africa. It appeared that, while here in Twiga, they met with an Australian medical team that had, until recently, been working in Ethiopia as part of a famine relief project and were now enjoying themselves in Kenya before deciding on what to do next. John had taken a keen interest in one of the nurses, Diana Shillabeer from the Adelaide hills, but she and the other nurse, Jane Mathews, and the doctor, Ian Stevens, had gone in their Landrover to visit the island of Lamu. As the road to Lamu was purported to be 4WD, Mike had borrowed a Landrover from one of the other campers to drive up while John had initially opted to stay behind. But, on second thoughts, he wanted to spend more time with Diana and was thinking about catching the bus north. I offered to take him pillion. He joyfully accepted and thus began a somewhat difficult journey.

CHAPTER TWENTY-TWO

Ancient Cities

John had never been on a bike before and did not have a crash helmet or any other motorcycle gear, therefore was just riding in jeans and shirt. The road was rough and we hit several bumps and potholes and, every time we did, he grabbed hold of me – by the shoulders! You can imagine how this affected my steering. I had to stop and have serious words with him instructing him to either hold onto me round the waist or, preferably, just hold on to the seat.

At this point I have to include John's perceptions of this journey given to me from his diary at a much later date:

'June 17th 1974

Once again I said goodbye to Loic (*the Frenchman who had been picked up by John and Mike in Zaire*) having dropped him off on the road to Mombasa. I drove the van to a hotel car-park and tipped an African to guard it and then jumped on the back of the bike, a 600 BMW (*actually it was a 500*). For the first 70 miles the road was fast but sandy and as it rained and sunned alternately I was soaked through and dried out a number of times. The sandy road made quite a handful of the two-wheeled machine but this didn't deter Linda who very confidently pushed on at speeds up to 70mph (*What!?*) while I hung on for dear life! The ferry at Garson was being hauled across as we arrived and, after we had boarded, it waited about five minutes while a herd of about a hundred cattle swam across the swiftly-flowing river, being swept downstream by the current. The herdsman swam with the leading cattle guiding them to the further bank. For about five miles after the ferry crossing the road was rather muddy but would have been easily passable in our van. However the bike carrying Linda and me, was a new and unforgettable experience, one I wouldn't have missed for anything. The slippery wooden bridges were rather treacherous and on more than one occasion we had picked a nice soft looking spot to aim for when we parted company with the machine. However, Linda very skilfully regained control. Just ten miles short of the quay we were pounding along, me with my head hidden behind Linda's back, eyes closed, almost asleep when she finally slowed down. I looked up to see a herd of elephants crossing the road about 200yards ahead. They were munching the roadside grass and, on seeing us, ambled slowly into the dense scrubland on either side of the road. When they had all left the road, we continued and, to our amazement, as we passed the spot where they had been we found, on looking into the scrub, no sign of them. They were so completely camouflaged.' (*end of John's version of our ride*).

Lamu town is on an island so we had to leave the bike on the mainland next to the two Landrovers, already parked, and take a ferry over. As it happened, the two girls were there beside the Landrovers and I was introduced to Jane and Shilly. Jane was an elegant blond and Shilly a very attractive brunette. I could see John's attraction to her. Both girls were friendly and welcoming. John and I took the ferry and the girls came later. They told us of the guest house that they had already booked into and John went to find the boys.

I wandered through the narrow streets between white-washed, flat-roofed buildings which had rooms to rent for the thriving tourist trade and shops selling shell necklaces and Ethiopian silver and jade. There were beautifully-carved wooden doorways on many of the old buildings and an interesting old fort near the central market. I stayed overnight in a tiny room with the other girls but could go up to the roof and see the graceful dhows bobbing in the blue waters of the Indian ocean and imagine the hustle and bustle of the times when this was a busy port.

Lamu is Kenya's oldest inhabited town. It was founded in 1370 by Swahili settlers. In 1505 it was invaded by the Portuguese but the inhabitants received aid from Oman to resist Portugal. Its golden age was between the late 17th Century to the early 19th Century and it became a centre of poetry, politics, the arts and crafts. It was considered a literary and scholastic centre and women writers received a much higher status here than in other parts of Kenya. At that time the economy was based on the slave trade (for which Lamu was an important port of export) as well as for ivory from elephant, rhino horn, turtle shells and mangrove.

Traditionally mangrove was used for charcoal and firewood but it was also used for the construction of dwellings, boats and fishing gear, tannins for dyeing and leather production, the latter uses being particularly followed in Arabic countries. Also extracts from mangroves were used in folkloric medicine as insecticides and piscicides and these practices continue to this day.

The slave trade was abolished in 1907. The British had already established a railhead at Mombasa so trade at Lamu diminished. The mix of cultures was obvious in the differing types of facial features and skin colours that I noticed in the population.

I only stayed one night as I was worried about my bike parked on the mainland so reluctantly left this fascinating island and started my journey back to Pole Pole. John was able to return in comfort in one of the Landrovers and I took the time to find some ruins I had heard about near Malindi. These were the Gede ruins, the remains of a mysterious pre-Portuguese Arab-Swahili town of about 2,500, which was abandoned in the 16th Century. The remains of the buildings were set in dense jungle over several hectares and were hard to find in the undergrowth over which towered huge boab trees. The whole place had a creepy feel about it. There were remains of a palace, several mosques and single-storeyed houses as well as many large tombs topped by phallic pillars. The architecture is Arabic but artefacts found there indicate trade with India, Persia and China. Malindi's ancient records mention Chinese traders who took slaves and ivory from along the coast and tell of the visits between 1417 and 1419 of the Chinese Emperor Cheng Ho.

While wandering around these fascinating ruins it occurred to me that, though I had been educated in a well-reputed English grammar school with subjects that included Geography and History, our ego-centric British education had not mentioned any of these amazing old civilisations. Africa and its people were depicted as ignorant and undernourished. How much we can learn by travelling and seeing for ourselves.

During the last hundred miles of my journey south, the bike's engine started making worse noises than usual. It had been rumbling for some time and I discovered to my horror that it had run out of oil. The main bearings were shot. I slowly rode back to Miki's to think about how to fix it.

In the town of Mombasa I had seen another large capacity motorcycle, a Honda 500/4 and found that it belonged to an American Peace Corp worker, Roger, who taught at the local technical college. He had heard the rumblings of the BMW engine and said that, if I needed a hand to fix it, to contact him so I had taken his address and kept the offer in mind. So now it was time to take him up on it and I was just about to do just that when the dreaded malaria struck.

Chapter Twenty-Three

An Enforced Stay

I suddenly developed a high fever, dizziness and the tremors, then collapsed. Miki had a little car and took me to a doctor in Mombasa. He carried me into the surgery where a number of African people were waiting and I fleetingly glanced several pairs of eyes watching curiously as Miki pushed past shouting, "A memsahib, very ill!" and immediately jumped the queue. He burst into the doctor's consulting room and laid me on the bed.

The Indian doctor took one look at me and said "Malaria. Have you been taking the anti-malaria tablets?"

"No," I admitted, "I ran out and didn't bother to buy more."

He took my temperature, which was 104F and prescribed quinine and more tablets.

"You must take these tablets and rest, there is nothing else to do until the fever passes. Keep up the fluids". I had visions of large glasses of gin and tonic but he meant water.

There are several different types of malaria and the worst is cerebral malaria which is very often fatal. No effective vaccine exists and the only way to try and prevent the disease is taking chloroquine tablets but these are not always effective. There is an inaccurate belief that, once you have had malaria, it recurs but this is only true if the treatment has not been successful and some parasites survive in the blood stream or lie dormant in the liver cells. (I found out later that Eion, the Irish guy who I had met in Kano, Nigeria, had used this mistaken belief about the recurrence as an excuse to take time off work once he had returned to his teaching post in Ireland).

It is advisable to use a strong insect repellent such as DEET, mosquito nets while sleeping and to burn the green anti-mosquito coils. In certain countries DDT is still used to spray large land areas and especially to control breeding of mosquitoes in stretches of open water.

Miki took me back home and, bypassing my tent, installed me in his house, in his own bed and nursed me himself. For a week I lay suffering with an intense headache while he gave me water and the little food that I could eat. I could hardly sit up without my head swimming and had to just lie there, alternatively shivering with cold and sweating with fever. He was most attentive, regularly taking my temperature and changing the sheets and I was very fortunate that he was so good to me. Finally I could get out of bed but could not walk for a week, and spent the time just sitting on the veranda and taking in the scene, reading and sipping cool water. But I was worried about my bike and was determined to take it into town, find Roger and repair it.

When I could get up I didn't have the strength to kick start the bike but asked Miki to do that and help me get astride, then I rode gently into town. Roger, a laconic string of a man with large glasses (fashionable at that time), was surprised to see me, especially in such a state but invited me to stay in his basic accommodation and we began stripping down the bike. As he was teaching in a technical college during the day, we worked on it in the evenings. Fortunately I had my BMW manual with me and Roger would sit studying it, after dinner, while smoking his evening joint. Then we began working in a back room with the green mossie coils burning around us. Attracted by the light, there were also a large number of huge brown moths fluttering in the room.

I was still very weak and it was all I could do just to sit on the floor and wash the engine parts in petrol as he dismantled them. To take out the crankshaft and bearings needed special BMW tools but Roger, a skilled engineer, designed some variations and, with the facilities available to him at the technical college, was able to make up some that would do the job. As he worked he talked about the latest news that was rocking the American public and that he had read about in the *Time* magazine which the expat Americans closely followed. It was the Watergate Affair and Richard Nixon was in the hot-seat and eventually had to resign. It all went a bit over my head (which was still spinning anyway); politics never really interested me much, only when it stopped me travelling to certain countries.

We found that the slinger rings, which collect the debris in the oil, were packed solid and needed cleaning out. The oil inlets had become blocked because of this and it had not been able to get to the main

INTO AFRICA – WITH A SMILE

bearings, thus they were completely knackered. Later model BMWs have oil filters which eliminate this problem and they also have different bearings. Luckily, on these old models the bearings are normal roller ones, 6207, and they are used for many different types of vehicles and machinery and were easily available in Mombasa. So we located some, fitted them and the engine was reassembled and ran as good as new.

I was so grateful to Roger. How lucky I was to meet him! He was so good to spend the time helping me even while he was busy writing end-of-term exam papers. I was also extremely fortunate that, while in town, I met two German riders who had travelled up from Johannesburg intending to go north but for some reason were not and, therefore, did not need the off-road tyre that they had brought with them. Fortunately it was an 18" and they were happy to sell it to me as mine was just about bald. I am still constantly amazed that, when travelling, people help one another and things always work out.

Needless to say I bought Roger a whole load of beer. I also wanted to take some back to Miki for his kind care of me and to wish him farewell as I was intending to leave Pole Pole and go out to Twiga Lodge to meet the Welsh boys as we had talked about driving down through Zambia together. So, I loaded my leather panniers with beer bottles for us to have a farewell drink. My bike might have had new engine parts but its brakes were so worn that I had almost no stopping power and when a car suddenly came to a halt in front of me I had to throw the bike down to avoid hitting it. I knew the bike would be OK but I was worried about the beer! Fortunately only one bottle was broken. We had a good party and sing-song with Miki playing his new lagerphone, and I fondly bade farewell to someone who had been exceptionally good to me.

Linda Bootherstone
Zambia to Zimbabwe

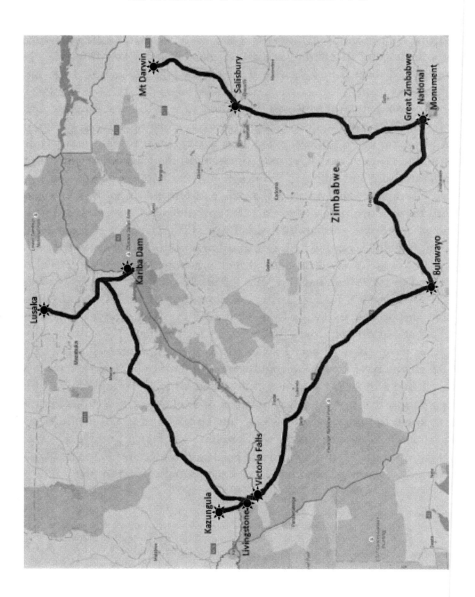

CHAPTER TWENTY-FOUR

Into Zambia

Leaving Pole Pole I travelled north once more to Twiga Lodge. The tourist resort was a guest-house-type place with individual huts dotted along the beach in one direction and a small, very pleasant campsite at the other. It was the off-season for tourists but I had been warned that one had to be careful as it had a reputation for robbers. Some backpackers, camped alongside the glorious, white, sandy beach had woken in the morning to find their tents slit and clothes and valuables taken – even from under their pillows. It was wise to sleep with your money belt on, which I always did.

John and Mike were based on the beach but Mike had gone on a trip with some others and wasn't expected back for a couple of days so we had to wait. This wasn't too bad as the beach was good, the weather very pleasant and there was a local bar. When I asked about the whereabouts of the Australian contingent it appeared that Diana (Shilly), Jane and Ian were going to visit some game parks and then Ian was driving to Nairobi to sell the Landrover before returning to Australia. The two girls were going to hitchhike south to meet the Welsh boys in Mumbwe near Lusaka.

Mike was rather a long time coming and John and I were just starting to get worried when he arrived with some Irish girls he had met in tow. Needing to get provisions for the trip we all drove back to Mombasa to go to the market there.

Mombasa is a port of legend, full of noise, colour and activity and it has a unique character created by its mix of many cultures. Like Lamu it had seen many traders and travellers. Its first recorded mention dates from the 12th Century and it was described by the Arabs as the home of the 'black Africans'. It was well known as a port and a trading centre that shipped ivory, skins and slaves to Arabia. The Portuguese arrived in the 16th and, over the centuries the Indian, Arab, Asian and European traders settled here. Toward the end of the 19th Century the British, in an effort to suppress the slave trade, established a colony.

In the big, bustling market was an array of exotic fruit and vegetables and all kinds of seafood, spices and trinkets. A variety of African and Indian people greeted each other dressed in their colourful kangas, sarongs or saris. The Swahili greetings rang out "Jambo-habari ?". "Misuri". I loved the sounds of this language and the Pidgin English and all the sights and smells of this area with its diverse mix of people. Finally we were ready to leave. It was now July and I realised that I had been on the road for six months. What a number of countries and amazing sights I had seen! And there was still more to come.

Leaving Mombasa, we made a slight detour to visit some friends of John and Mike's along the coast and then, dropping off the Irish girls to hitch-hike back to Mombasa, we continued south. I had a bit of a worry when the bike started making terrible noises and I alerted the boys to the problem. On investigation we found that the nuts on the base of one of the cylinders had come loose, causing the barrel to move with the piston and this was causing the noise. It was simple to re-tighten them and, fortunately no damage had been incurred and the bike went perfectly afterwards.

We travelled through Tanzania on its mainly flat open tundra plains with only a small stretch of dirt road and made the trip as fast as possible to Lusaka in Zambia. The border crossings Kenya-Tanzania-Zambia were easily accomplished but we were somewhat hampered in our progress by the numerous blow-outs that the Combi suffered. The tyres were completely bald; long past their use by date. We had to patch them up with our rubber thongs and used branches of trees as a jack, as theirs had long since broken.

Another excerpt from one of John's diaries describes one of the incidents along the road.

"July 30th 1974

In Morogoro we found a garage to fix two of our punctures while we had a meal. Getting out of town was a bit difficult as signposts are a rare find but eventually we were on the road again. By now it was quite warm inside the van despite wide-open windows. Linda's bike was performing well and we made good progress.

Through the Mikumi Game Park we stuck close together as there were elephants grazing on the road-side. Linda was quite scared on her bike. (She had been prevented from travelling through Serengeti because of the danger). As we rounded one corner, with Linda about 20 yards behind, I saw a huge snake with head reared, in the middle of the road. It looked like a cobra. I had to kill it with the van, much as I didn't want to, as it could certainly have given Linda sufficient fright to throw her off her bike. After Nikumi town we climbed into some wonderful mountain scenery, green thick foliage, a pleasant change after the dry, flat plains we had crossed. We travelled well on the sealed, fast roads but it was somewhat disconcerting to meet all the trucks transporting copper to the rail heads. Unfortunately they liked overtaking each other without making sure the road was clear and many had head-on crashes. The evidence of these, many burnt out vehicles, were everywhere along the roadside. We had to be very careful overtaking these careless drivers ourselves."

There was much evidence of Chinese influence in both Tanzania and Zambia. They were building a railway from the copper belt in the west to the port of Daar es Salaam. We weren't sure whether it was to get the copper out or arms in, as there was much unrest amongst these southern African countries, recently made independent and with their smouldering tribal differences.

In the late 19th Century Tanzania had been part of German East Africa but post-WWI, it was a British mandate and became independent in 1962. From 1970-75 it was aligned with China, thus they were helping with the railway. This rail route was important as it would lessen the necessity of sending the copper and other goods to the southern ports on the established railway to South Africa.

Zambia was originally the home of the Khoisan people – the Bushmen- but when the Bantus moved in they went further west into what is now Namibia. Zambia is named after the Zambezi River which means God's river. The earliest Europeans to visit were the Portuguese then David Livingstone came and made his discovery of Victoria Falls in 1855. Cecil Rhodes, the British entrepreneur, set up his British South African (BSA)company, obtained mineral rights and sent his scouts to the northern areas to look for minerals and ways to improve navigation of the rivers. He found copper deposits along the Kafue River in what was then called Northern Rhodesia (after Rhodes who had so much influence amongst both whites and blacks in the area.) In 1923 BSA ceded control of Northern and Southern Rhodesia to the British Government and finally in 1964 Tanzania became independent under the Prime Minister, Kenneth Kaunda.

The black African countries were increasingly against the white regime in Southern Rhodesia and South Africa and Kaunda endorsed the guerrilla raids into Southern Rhodesia. Political tension and militarisation led to the border between Zambia and Rhodesia becoming closed in 1973. This indeed affected our travel.

Fifty miles north of Lusaka we came across an old English-style pub called *The Ploughman's Arms*, apparently very popular with the colonials before independence. Now it was very run down, lacking its former glory. The paintwork was chipped and faded and the furnishings were worn and tatty. However, not to look a gift pub in the mouth, we stopped for a couple of hours for a celebratory beer or two as it was only a short way to the capital.

On arrival at Lusaka we made immediately for the Post Office to check our mail. As with most African Poste Restante, it was necessary to sort through <u>all</u> the mail, not just the pile with the alphabetical letter of our name, as the African sorters couldn't always read. Mistakes often occurred. I found quite a few letters for me including one from my mother with some sad news.

I had become involved with motorcycling because I had fancied my older sister's boyfriend, Dave Nixon, who lived in the next street and owned a lovely BSA Gold Star motorbike. These were considered 'the bees knees' of bikes at the time (together with Triumph Bonnevilles) and, though Dave was a skinny, spotted being with a squeaky voice that had never fully broken, he was considered quite a catch. We called him Big Boots and the pop song at the time 'I want to be Bobby's girl' was changed to 'Big Boots girl'. After he split with my sister, I shamelessly chased him but he didn't fancy me and I resigned myself (well, not quite) to being just a friend. He became involved with production racing using Triumphs and I accompanied him when he went with other local boys: Barry Tingley, Colin Agate, Paul Coombes, Ray Knight and more, to production-race meetings at Brands Hatch, Snetterton and Silverstone. They used Ford Thames vans to transport their bikes and there were often impromptu races between the vans on roads such as the A11 or A2, on their way to the circuits. It was all good fun and, though competitors, they always helped each other out in the pits if some-one needed an extra sprocket or tyre etc.

Dave went on to become a sponsored rider in the Isle of Man. He married and had a child, I believe. However, in that year of 1974,

Into Africa – With a Smile

he was killed while racing on the Isle of Man. I was very upset to hear about this. Over the years a few of my friends had been killed while riding either racing or on the road but it was a sport and interest that they enjoyed so, what better way to go?

After another beer to wash down the bad news, we spent a night in Lusaka and continued the next day to find some friends of John and Mike's who owned a copper mine.

CHAPTER TWENTY-FIVE

Hippo Mine

Turning west off the main road to head toward Mumbwe in the Kafue National Park, we were now entering a tsetse-fly zone of which we soon became aware. As we were travelling painfully slowly along a dirt track, I saw a small hut ahead at the side of the road with a stop sign beside it. Dutifully coming to a halt, I sat still while a local holding a huge fly spray (the old Flit type with a tank and pump handle) came out of the hut, approached and started to spray me and the bike all over with what I presumed was something like DDT. I quickly shut my eyes and mouth. I made a querying motion and he pointed to a sign which did in fact say that it was a tsetse-fly post. He moved on to spray the Combi behind me.

Out in the open at a slow speed, I was indeed beginning to be bothered by the large, stinging flies around and it was probably just as well I had been sprayed. Tsetse-flies, though looking similar to house flies, are a pest in mid-continental Africa as they carry the disease of sleeping sickness (African trypanosomiasis). The symptoms are swollen lymph nodes, fevers, headaches, joint pains, confusion and poor co-ordination and without treatment can lead to cardiac and kidney dis-functions. They caused thousands of deaths per year in Africa in the 1970s. Having just had quite a few of those types of symptoms with malaria, I was not keen to have any more.

We finally arrived at the homestead of the Hippo Mine owned by Mr and Mrs Kropechech. Vladislav was from Czechoslovakia and his wife, Gwyneth, was from Wales and was actually a friend of John's parents. They had been expecting the boys for some time. Shilly and Jane had arrived previously and Gwyneth was delighted, it wasn't often that she had the chance to chat with other white women. They installed us in a rondavel, a little white, round house made of mud and whitewashed in and out. We set up our sleeping gear inside and two smaller rondavels nearby were set up as a toilet and bathroom.

116

Luxury!

We spent a week in this interesting area with Vlad explaining the workings of the mine. Although it was in the national park, he had a permit to shoot game to feed his workers so one day we went out with him and his son to procure some meat for food. We formed a line with Vlad leading, gun at the ready while the boys followed and we girls took up the rear. From this outing we came back triumphant as Vlad had bagged a kudu and we had a BBQ that night. Gwyneth had a piano and, being Welsh, was fond of singing so we spent many happy musical evenings. With John and Mike joining in, it was just like having my own Welsh choir.

Vlad's right hand man was a black Rhodesian called Edmond who was well educated and had travelled extensively in Africa. We spent many interesting evenings with him and his wife too.

One day I was horrified to discover that I had lice in my long hair. Embarrassed, I asked Gwyneth what I should do.

"Go into the bathroom and run a bath. Get into it and I will be with you shortly," she instructed. This I did and she arrived with a bottle of kerosene.

"Now keep your eyes closed while I rub this into your scalp."

I screwed up my eyes tightly and tried not to wince when the stinging liquid was applied. After a couple of minutes she said, "Now put your head under the water and give it a good wash with this soap." She handed me a bar of something similar to Lifebouy.

Well, it worked. The little horrors and their eggs were washed down the plug-hole and my hair was once more lice free – though my scalp was rather sore. A very effective treatment and one that Gwyneth used often on her local 'family'. She told us that she also had a big supply of first aid materials as, every pay-day, the locals imbibed their 'pombe' made from fermented corn or fruit. Mealie meal, a corn porridge, was a staple diet of the people of much of southern Africa. It was soon discovered that this could be used to make an alcoholic brew. It looked like watery porridge and I did try some but, to me, it tasted disgusting. Both men and women partook of this potent brew and the results of their intoxication were in many cases serious. Fights broke out and large sticks and sometimes pangas came into play. Gwyneth spent the next morning bandaging up the participants and expressed amazement at the damage incurred and how some, with horrendous head injuries, were still alive! Vlad said that their skulls must be super thick as "these injuries would have killed a white-fella". However, once bandaged up and given an aspirin the patients were quite content, held no grudges and happily lived together again – until the next pay day party.

While we were there it was over a public holiday and so the mine was not in action but Vlad took the boys on a tour of it anyway. John later described the frightening entrance to the mine which consisted of a rope ladder hanging down into a dark cavern. He said he was 'shit scared' while using it. Apparently the digging was all done by hand with pick axe and shovels and the ore also pulled up by rope; a very primitive form of mining.

Although we were enjoying our stay with these kind people, Jane and I were getting itchy feet and wanted to be on our way so I agreed to take her pillion down to Rhodesia. In re-loading the bike to make room for her, I picked up my tent that I had carelessly left on the ground outside the rondavel. To my horror I found that a group of white ants had made a meal of it and it was only just useable. We took off, with Vlad escorting us along the dirt road for some way and it was just as well he did for I had a puncture and he helped me replace the inner tube. Finally we made it out onto the bitumen and the road to Livingstone.

CHAPTER TWENTY-SIX

The Road to Victoria Falls

Jane had never been on a bike before and she had no crash helmet so she just used a scarf. Although her shapely bottom was unaccustomed to miles on a pillion seat, she did not complain and, now on sealed roads, we continued for 200 miles without even stopping for a lunch break or to stretch our legs. While working in Ethiopia, Jane had accumulated some lovely silver and jade jewellery which she carried by wearing it. She cut quite an elegant figure with her slim body, blonde hair and blue eyes adorned with necklaces and bracelets. Quite a contrast to me in my tatty clothes!

About 50 miles north of Livingstone we were stopped by local police at a check point. The officer wanted to see my licence and to make sure that my indicators and horn were working. Being a 1957 model the bike had not been fitted with indicators and the horn hadn't worked since the button had been smashed when it fell over on the ship. The officer didn't like this at all and wanted to fine me. We explained the indicators and feigned surprise that the horn was faulty saying that we would get it fixed immediately in Livingstone. Eventually, after much talking, he let us go and, breathing huge sighs of relief, we continued on our way little knowing that worse was to come.

In Livingstone we were stopped at another police check point and I drew up expecting problems with the horn again but the officer asked for my driving licence and, on inspecting it, declared that it was forged! Well, strictly speaking it was. When I had received this International driving permit from the Automobile Association in the UK in January that year they had put on the wrong date stamp. They had forgotten to change it from 1973 to 1974. With not enough time for me to go back to the London office to get it corrected I had simply changed the date myself. No one had previously challenged this but this officer was obviously looking for promotion. He said my licence was out of date, I'd changed it and would be charged with forgery. He also said it was illegal for Jane to ride without a crash helmet and, looking around for something else to find amiss, he decided that, since I didn't have a Zambian number plate, I was uninsured. I tried to explain about my International Carnet de Passage but he was insisting that we would have to go to court and my bike would be impounded. Quite a crowd had gathered and he was obviously pumped up with importance. I was just about in tears; we had come so far and the Rhodesian border and sanity (at that time) was just a few miles away and this clown was going to lock us up!

Jane, with a much cooler head than me (it wasn't her bike that was going to be taken) smiled sweetly at the officer, batted her long eyelashes over her big, blue eyes and said, "But Officer, I really don't think you need to worry about us as we are going directly to Botswana, out of your area. We would be eternally grateful to your good self if you would let us continue to that country. I am a nurse and am going to work there." It was de riguer not to mention Rhodesia or South Africa but Botswana was a fellow black nation and Jane's story proved to be the magic words.

Suddenly he said, "In that case, Madam, you may continue."

I was a complete nervous wreck by this time and couldn't get the bike started and away fast enough. We were both still shaking and laughing with relief half an hour later in the nearest pub.

Feeling suitably refreshed and fortified by a beer, we made our way to the great Zambezi River and the northern side of the legendary Victoria Falls, first brought to the attention of the world by David Livingstone in 1855. The militarised zone meant we were unable to get a good view but gazed longingly at the road/rail bridge that spanned the river and marked the Zambia, Rhodesian border. Unfortunately, due to the political rift between the two countries, the bridge and border had been closed the previous year and we would be unable to cross it. Instead of a 100-yard skip across the river, we would have to make a detour of nearly 200 miles through Botswana which still had an open border with both countries. That night though, while camped nearby, we heard the rumble of train carriages being shunted to the centre of the bridge and there linked up to Rhodesian engines to be surreptitiously taken into this 'illegal' country. In spite of politics, it seems trade must go on!

Into Africa – With a Smile

Pole Pole   Miki

Visiting Miki's friends

Miki and friends at ruins

Lamu

Lamu street

Linda in Kenya

Linda Bootherstone

Chair in the Lamu Museum

Dhow in Lamu Museum

Lamu Dhow

Lamu seafront

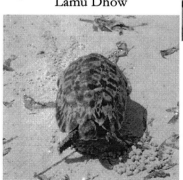

Turtle

Cows on the beach

Shilly and John

Into Africa – With a Smile

Mount Kilimanjaro

Barry Tingley

Meerschaum Pipe

Dave Nixon

Jane Matthews

VICTORIA FALLS

Bridge across the Zambezi

Victoria Falls

Above: Tourists on the path to the Falls
Right: A rainbow in the mist

CHAPTER TWENTY SEVEN

Rhodesia

So, what exactly was going on in Rhodesia at this time?

Like Zambia, Rhodesia had been under British rule since the 1800s but in the 1960s, while other countries were given independence, the white government of Southern Rhodesia, under Prime Minister Ian Smith, refused to hand over to black rule. The reason was that the cabinet of Rhodesia disagreed with the British Government over the terms under which Rhodesia would become independent. The white settlers were afraid that the black politicians would not govern it properly and that a Congo-style situation may result. In 1965 stalemated talks between Ian Smith and British Prime Minster, Harold Wilson led to Rhodesia declaring a Unilateral Declaration of Independence (UDI). Britain was somewhat taken aback by this; it was the first breach by one of its colonies since the United States Declaration of Independence 200 years beforehand. The Smith administration initially professed loyalty to the Queen but abandoned this in 1970 when Rhodesia declared itself a Republic in an unsuccessful attempt to win foreign recognition.

In this year of 1974 it was still stalemate and Rhodesia was fighting a bush war with the African National Union (ANU) and the Zimbabwe African Peoples Union (ZAPU). We were effectively about to enter a country under siege and with internal terrorist activity. Most countries, apart from South Africa, had now issued sanctions against Rhodesia and therefore trade between them had supposedly been stopped. The trains on the bridge indicated that there were still some unsanctioned dealings occurring – where there's a will there's a way.

And so it was for us. We battled 70 miles of dirt road east to reach the rickety Kazungula ferry which crosses where the Chobe River meets the Zambezi over a 400m wide stretch of water. The town of Kasune on the other side is a point of junction of four countries – Namibia, Zambia, Rhodesia (now Zimbawe) and Botswana. On entering Botswana, I had to go to all the trouble and expense of getting insurance for that country even though, just a few miles up the road, we entered Rhodesia. Oh, what a joy to see white, smiling faces and be greeted by "Welcome to Rhodesia" and have our passports and carnet stamped by efficient, neatly-dressed officers. We had no problems getting through immigration as Jane had plenty of money to show and I had a return ticket to England. We both produced a few address of 'relatives' in Rhodesia who would accommodate us. I had to laugh when we were warned of the 'bad' road to Victoria Falls which was only 40 miles and took us 50 minutes. After what I'd been through it was a highway!

When we reached the hotel on the southern side of the falls we again took a walk around and were able to admire the thundering water over the steep drop and the myriad of rainbows that it throws into the air. It certainly is a magnificent sight and must have been quite a discovery for David Livingstone when he came across it. He named it in honour of Queen Victoria, but the indigenous name, Mosi-oa-Tunya –"The smoke that thunders"- continues in common usage as well. On its way to the falls the Zambezi river flows in a shallow valley. There are no mountains or escarpments only a flat plateau. Just at the crest of the falls are two islands that are large enough to divide the curtain of water, even at full flood: Boaruka (or Cataract) Island and Livingstone Island, so named as it was the point from which Livingstone first viewed the falls. The falls are formed as the full width of the river plummets in a single vertical drop into a chasm and into the First Gorge then through a zigzagging series of gorges until a sharp turn has carved out a deep pool called the Boiling Pot. Objects – and humans- that are swept over the falls, including the occasional hippo or crocodile, are frequently found swirling about here or washed up in the Second Gorge. In 1910 two bodies were found here, mutilated by crocodiles after two canoes were capsized by a hippo on one of the islands above the falls.

There are walkways around the falls which are safer during the dryer season as in the wet they are shrouded in mist and spray. At full moon a "moonbow" can be seen instead of the usual rainbow.

Though neither the widest or highest waterfall in the world, it is classified as the largest based on its width and height. It is the world's largest sheet of falling water. Roughly twice the height of North

America's Niagara Falls it is, to my mind, far more romantic and definitely less commercialised.

While having a celebratory beer at the hotel we were saddened to hear that only a few weeks before a tourist had been shot by a sniper located on the Zambian side.

After a very good night's sleep at the campsite, we continued in high spirits on the road to Bulawayo. Only 40 miles out and there was a loud bang - a noise I knew only too well from a previous experience in Australia a few years before – the sound of a valve dropping through a piston. This time I knew to stop immediately, even if the bike was still running on the other cylinder. Poor Jane, her first motorcycle trip and she certainly was experiencing the highs and lows.

There was nothing that I could do to fix it there, so we were waving down passing vehicles. The bike was so big no-one could help us, although there was a lot of tut-tutting and head shaking over the damage. Eventually an army truck with about 10 guys in full uniform stopped. The sergeant/lieutenant and all the soldiers jumped out and there was lots more tut-tutting. The man in charge said he couldn't actually take the bike and us to Bulawayo because it was against the rules to take civilians but they would stay until someone came along who could. Sure enough an empty cattle truck soon appeared. The sergeant waved it down and all the soldiers lifted up the bike and put it on the cattle truck. I sat with it, in the back amidst the cow shit, to make sure that it was stable. When the driver stopped for lunch, I decided that it was stable enough and joined him and Jane in the front cab though I didn't smell too good by this time. The driver took us to John Lowe motors, a car dealer. The African boys there lifted the bike down and the white receptionist was kind enough to make us a cup of coffee which we gratefully drank while she phoned the local motorcycle dealer, Geoff Lacey. He arrived shortly and took us and the bike in his ute (termed 'bakkie' in Rhodesian slang) to his workshop. We booked in at the YHA and stayed there a couple of nights together before Jane managed to get a lift to Salisbury while I busied myself stripping the top end of the damaged cylinder to make repairs.

Chapter Twenty-Eight

Rhodesia's Foundation

Luckily for me a young English couple, Del and Glyn, who had recently emigrated to Rhodesia, were happy to accommodate me at their house where we could reminisce about England and the bike scene there. Bulawayo, although important as a railhead, was a small town and I could easily walk to the bike shop where Geoff didn't mind me working on the bike myself and lent me any extra tools that I needed. Looking at the damage that the dropped valve had caused, I found that I needed a new piston and rings and, although I had a spare exhaust valve with me, was lacking in the retaining collets. Geoff rang around known BMW owners in Gwelo and Salisbury to see who may have spare parts to help.

It appeared that my visit to the hallowed hall of the BMW club in London had paid off. The members there had indeed put an article in their magazine about my intended ride through Africa and the Rhodesian and South African contingent were expecting me. From both Salisbury and Gwelo, parcels arrived and amongst the parts we found an oversize piston that fitted but had to use car rings. No BMW valve collets were available so we used some from a Matchless. These did the job and within a week the bike was up and running.

The motorcyclists in Bulawayo were really friendly and I took great delight in their company. All the way through Africa I had been the only motorcyclist travelling south. Ken had just used his bike as transport and hadn't enjoyed it. Roger in Mombasa was an enthusiast but not travelling and the Germans who had sold me their tyre were off in a different direction. Now I was with people who truly shared my interest and we went for a ride together to the Cecil Rhodes grave at 'World's View' in the Matopos ranges. The countryside was really pretty and the view from the hilltop, where he is buried, showed a landscape dotted with trees and kopjes. During the week I spent three afternoons in the museum in town which explained the history

INTO AFRICA – WITH A SMILE

of Rhodesia and Cecil Rhodes, a man very much of his time and place.

Born in Hertford, UK in 1853 he was the son of a Church of England vicar. A sickly child he was sent, as a teenager, to South Africa to join his elder brother as a planter. It was thought that the sea air on the voyage and the better climate in Natal would improve his health. Cecil and his brother, Herbert, failed as cotton planters and moved on to try diamond mining in the Kimberley. Cecil, using borrowed money, invested wisely and began accumulating wealth. He became a partner in De Beers diamond mines and also in the Niger Oil Company. At a later date he also became involved in fruit farming in the Cape Province.

For a short time he returned to the UK to study at Oriel College, Oxford. There, meeting many members of the ruling classes, he became attached to the cause of British Imperialism. He truly believed that Englishmen were of superior quality and should rule the world. It was, of course, when the British Empire was at its strongest. On returning to South Africa he entered public life and, in 1890, became Prime Minister of the Cape Colony. Although he had many disagreements with the Transvaal government policies, he had respect for the Boer work ethic. During his tenure in the Cape Province he introduced an act to push black people off their land to make way for industrial development. Using his wealth and contacts he set out to conquer the African territories to put them under British rule although he wanted to administer them himself rather than answer to the government in London. In drawing up a treaty that the Matabele king, Lobengula, did not fully understand, he cheated him out of his lands and, when the Matabele and Mashona rebelled against the coming of white settlers, Rhode's company police crushed them. Strangely, during one battle Rhodes risked his life by negotiating unarmed with the chiefs; an act which won him their grudging respect.

His company, The British Central Africa Commission, wanted to start a new 'Rand' (named after the gold mining area in the Witwatersrand near Johannesburg) in the northern part of South Africa and encouraged white settlers to move there but, as gold was not so evident, they began farming instead. In 1895, because of his heavy involvement and massive investment, these territories were named after him thus Northern Rhodesia – north of the Zambezi and Southern Rhodesia, south. (now named Zambia and Zimbabwe).

Rhodes truly believed that the British Empire should be worldwide and that English government was the best way to ensure peace amongst diverse people. He instigated the Rhodes scholarship for students, from British territories (current and former-such as the USA) and also included Germans whose work ethic he admired. He was a very hardworking and controversial man who had friends and enemies, both black and white, in high places. He died at the early age of 48 years in South Africa but had willed that he should be buried at World's View. At his funeral a special train brought his body from Cape Town to Bulawayo and his coffin was accompanied by many black and white people to the grave. With such a strong predecessor initiating this country and its policies it was hardly surprising that Ian Smith had taken his own stand against the British government.

Finally I took leave of my new friends and followed the road to Salisbury, stopping off at Gwelo to meet and thank the BMW owners that had sent me the spares. The bike was making funny noises and had stopped charging but I pushed on to Salisbury with fingers crossed that it would make it to the next BMW owners for whom I had been given an address.

Chapter Twenty-Nine

Salisbury

Rob and Rose Rushforth, a couple in their early forties, owned a motorcycle shop and were well-known and liked in the town. Rob had built an outfit based on a BMW but had replaced the engine with that of a Volkswagen beetle. He had made such a professional job of it that it looked factory-built and was his pride and joy. They kindly invited me to stay in their comfortable home, complete with native maid, and Rob said he would take a look at my ailing bike after his holidays. In the meantime I found myself a job!

At that time in Rhodesia all the manual work was done by the natives and the office jobs were reserved for the whites. I simply went to the employment agency and put my name down. I found out very quickly how straight-laced and old-school the Rhodesian society was. My passport was still in my married name of Mrs Linda Bick but when at my first interview for an office job I told them I was separated, they looked at me with disapproval and I didn't get the position. I was quite upset by this and decided that I would approach things differently. So, at my next interview my story was that I had flown to Rhodesia alone at present as my husband was still winding up our business in the UK. He would join me as soon as he could. I passed muster and was employed as a receptionist in a lawyer's office in town. My cover story was blown when the local newspaper did an article on my lone trip down through Africa but I was kept on. The switchboard was a bit of a mystery to me, although I had said that I was familiar with that type. I had to learn its idiosyncrasies pretty quickly and a few connections were cut as I was doing so.

Rob and Rose kindly lent me a little 125 Honda for transport while my bike was off the road and I used this to ride to work. However, I was not allowed to wear trousers in the office so had to buy myself a dress and changed clothes on arrival. I was surprised at the long hours in offices, 8am-5pm, with just the hour for lunch and

I took a while to get into the routine. I felt so exhausted in the evenings that I wondered if I had caught bilharzia, but it seemed it was just the belated effects of my journey and I revived.

My fellow workers were friendly and helpful but I found their attitude to the Africans quite hard to take. They really believed that, not only were they a race apart but they were inferior beings and, to most, the thought of integration was abhorrent. "You wouldn't mate dogs with cats" was one comment that stuck in my mind. One day, while we were out riding, I noticed a building site for a dam. There were many Africans digging with spades and moving the earth in wheelbarrows. When I asked my friends why a front-end loader wasn't used for this work they replied that, if it was, the Africans would be out of a job. Their wages were obviously cheaper than the cost of machinery. Another African job was replacing the skittles when we knocked them down when playing at a Ten-Pin Bowling venue. Young black men squatted behind the alley to pick up the skittles and ready them for the next bowling ball. Sometimes I had to bite my tongue when listening to discussions on the 'native problem'.

But in all fairness given the extreme attitudes of their present and past politicians, it is hardly surprising that the general population had continued in this mind set. The Africans had not been given the opportunities to become educated and therefore appeared stupid to their employers. The whites had the knowledge and control of the land and it would take a long time to change. Like Ian Smith, many white Rhodesians had been born in the country and considered themselves as African as the Africans. They didn't want to lose what they had worked so hard for and really did not believe that the native people were capable of ruling the country – at least not to their satisfaction. Even Rob and Rose endorsed this attitude so, in order to widen my social circle and mix with some non-residents, I called in at the YHA to see if there were any other travellers around.

Salisbury was a very pleasant town and easy to get around. The buildings were mainly single-storeyed and there were many parks. The YHA was situated in a street that was lined with jacarandas and, as this was the season when they were flowering, their glorious purple blossoms were in full bloom and the pavements covered with fallen petals. Now, whenever I see jacarandas it reminds me of Salisbury for this was the first place I had seen such trees.

In the YHA I met a young man, John, from Adelaide. He was

INTO AFRICA – WITH A SMILE

visiting Rhodesia and South Africa to see the steam trains, as he was an enthusiast and was making notes of all the British-built engines that were still in use in these countries. We went out for a few drinks and rides together, me taking him pillion on my borrowed bike.

I was really enjoying my stay in Salisbury; there were plenty of people to go out with on rides and I loved exploring the area. Rob stripped down my bike and found that the crankshaft had bent when I dropped the valve. He straightened it out and replaced the dynamo brushes that had worn out due to the imbalance. Unfortunately there hadn't been any BMW piston rings the right size when we replaced the piston in Bulawayo and we'd had to use car rings so it was burning more oil than it should but still, it was a goer.

Because of UDI and the consequent break with Britain, I was expecting the people to be anti-English but the reverse was true. Many white Rhodesians considered England to be 'the old country' and were very happy that I had come. One day, while out riding my BMW a woman, spotting my UK plates waved me down and excitedly asked me where I was from in England.

"And do you need any money for your trip?"

"No, no, I'm fine but thank you so much for offering."

They certainly welcomed me to their country and were sad that Britain had not understood their point of view. They were mainly hardworking, honest and generous people, taking pride in their country. I heard later, that when Cyclone Tracey hit Darwin, Australia in December 1974, the people of the small town of Mt Darwin in Rhodesia, took a collection and sent aid money, feeling like they were also members of the Commonwealth.

UDI meant sanctions and therefore many items that had previously been supplied by the UK were no longer being imported. However, this had a positive side to it. While I was there they had the 'Salisbury Show' and it displayed livestock and agricultural machinery. Not able to import the foreign-made models, Rhodesia had started making their own as well as pot-bellied stoves and other products not previously manufactured in that country. I noted a new process for plastic coating motorcycle frames. The show was good fun with the usual fairy-floss, toffee apples, fun-fare rides for the kids and country-style music.

On the music front, I found a folk club in Salisbury and was soon performing there with my lagerphone, which always drew interest.

133

LINDA BOOTHERSTONE

Apart from singing Australian and English folk songs I was getting used to the southern African accents and learnt a few songs in Afrikaans. This language is derived from the early Dutch settlers and used by the Afrikaans population or Boers (originally meaning farmers). Although the Rhodesian population was mainly of British descent there were some Afrikaaners living there.

While I was in Salisbury I had a letter from Ken Tilley, my troublesome travelling companion in Central Africa. It appeared that he was now in Salisbury as well and had his previous job as a warden in the prison. He invited me there to meet him for a meal. I couldn't believe my eyes when I saw him; now several pounds heavier he was clean shaven and smartly dressed in his prison officer's uniform. Almost unrecognisable! It was good to see that he had finally made it, after some very rough travel on boats and buses and hitching rides. He still hadn't heard from Deirdre but at least now had a job and an income. He was happy in his work and I wished him well and was not to see him again until many years later.

Into Africa – With a Smile

Somewhere in South Africa

Glyn and Del, Bulawayo

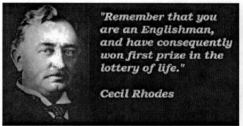

"Remember that you are an Englishman, and have consequently won first prize in the lottery of life."

Cecil Rhodes

Pioneer's grave

Del and Glyn in Matopos

With Rob and Rose in Salisbury

Racing at Salisbury

Rose at race meeting

Sidecar racers

Salisbury, Rhodesia

VW special in Salisbury

Ian Smith

Into Africa – With a Smile

Rhodesia bike run

Dassie rally

Riders, Dassie rally

Girl rider, Dassie rally

Winner of long distance award, Dassie rally

New bike

Big tank

Old bike

Zimbabwe site

Del and Glyn in the ruins

At the Zimbabwe ruins

Rallyists at Zimbabwe

Aloes at Zimbabwe

Left: Kariba Dam
Above: Rallyists at Kariba Dam

Rhodesia to South Africa

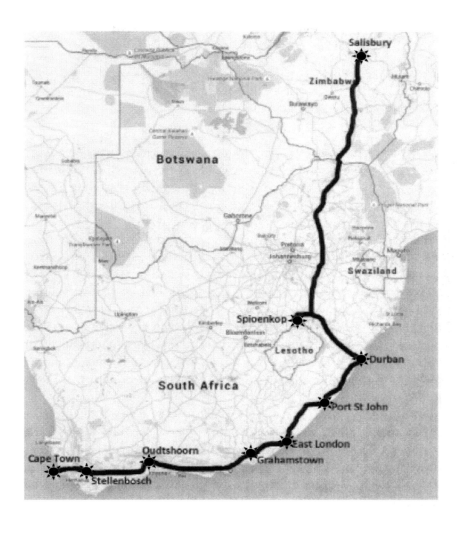

CHAPTER THIRTY

To South Africa

I spent two very pleasant months in Salisbury where Rob and Rose kindly let me stay with them while Rob repaired my bike. My receptionist job had enabled me to save enough money to pay for the repairs and to continue my journey so I braced myself for the move.

The Salisbury Motorcycle Club, of which Rob was president, organised an event called the Dassie Rally, held in the Kyle National Park north-east of the capital. I wanted to attend this before leaving and, in company with the local bikers, rode up to the rally-site.

There were about 200 participants from all over Rhodesia and South Africa and the long distance award was won by a girl from Botswana arriving on a 250 Honda. I was given a special award as a visitor. I may have travelled the furthest but not specifically that weekend for the rally. I was presented with a huge copper coloured bottle-opener with the inscription of the Rally and date on it. Very apt.

Several of us rode out to see the Zimbabwe ruins, which, in the early morning mist, derelict among the yakka trees, had a very eerie feel about them. There was no signposting or information plaques explaining them and, amongst our little group no-one had any idea who had built them or when. I heard that it was generally believed that this type of construction was too sophisticated for black Africans to have made but that is far from the truth.

Constructed by ancestors of the Shona people between 11th and 15th centuries the buildings served as a palace and seat of political power. The ruins are spread over about 7.3 square kilometres and are some of the oldest and largest structures located in southern Africa. They are of stone, without mortar and some of the walls are over 5 metres high. There are remains of three different complexes and it is suggested that they were either built at different times or used for differing purposes. The heyday of the area was between 1200 -1500

INTO AFRICA – WITH A SMILE

and it was a centre for trading in gold and ivory. It is suggested that its decline was due to less trade compared with sites further north, the exhaustion of gold mines in the area, political instability or even water shortage.

The bird design that has now become a symbol of Zimbabwe was inspired by the five soapstone bird sculptures that were initially installed on the walls and monoliths of the site. However they were 'souvenired' by European settlers and one was sold to Cecil Rhodes. The site was certainly impressive and at that time we did not know that its name would be given to the whole country.

Continuing our sightseeing tour we also visited the site of the Kariba Dam which was still under construction and rode through the town of Inyanga. Riding in areas close to the borders of Zambia and Mozambique we were told to look out for 'terrs', terrorists, members of the ANC who were fighting the 'bush war' with Rhodesia, pushing for black control.

After the Dassie rally I rode back to Salisbury to prepare for my leaving, bade farewell to Rob and Rose and my other Rhodesian friends and then left the town in company with some South African riders who had come north for the rally. It was good to have company and, as it transpired, very fortunate for me. Before leaving Salisbury, I had jokingly promised Rob that I'd wait until I left Rhodesia before I broke down again. We crossed the border into South Africa at Beit Bridge and just about 10 miles south there was another horrible noise and the bike came to an abrupt halt. It was obviously something serious, so, having a tow rope (who would ride without one?) we wound it round the steering column and I held on to the end with my handlebar grip while being towed by a 900 Kawasaki the 500 miles to Pretoria. The road was straight and flat and we were able to get up quite a speed, more than I usually did. Think of the petrol I was saving! We were a bit nervous of 'terrs' as they also took pot-shots of travellers along this route. Following along the river valleys while gradually increasing altitude, we made Pretoria in a day with no further troubles, apart from my aching wrist. One of the boys kindly let me stay with him and his wife in this capital city. Like Canberra in Australia it is the political seat of the country.

Club Motors of Pretoria was the BMW dealer and my bike was taken there for a strip down. It appeared that the piston, on the other

LINDA BOOTHERSTONE

side to the previous damage, had cracked and the bike needed a complete top end re-build on both cylinders, given the odd bits we had put it together with in Bulawayo and Salisbury. This took about a week during which time my host kindly lent me his wife's 400/4. I'm still not sure whether she had entirely agreed to that!

When presented with the bill at Club Motors I nearly had a heart attack but they kindly said that I only needed to pay for the parts as they donated workshop time in return for me having a photo taken outside their dealership to be put in the paper together with a write up about how kind they had been to this overland traveller. I had no problem with that! I was very grateful to everyone who had helped me on the ride and in Pretoria and, after my long tow by the 900, I looked at Kawasakis with new respect.

With a smoothly purring BMW I continued south to face the huge city of Johannesburg.

CHAPTER THIRTY ONE

The Lay of the Land

Perhaps at this stage it is pertinent for me to give an overview of South Africa's complex history. I certainly didn't know what was going on and how it had come to be one of the most disliked countries in the Western world. I knew it was because of Apartheid, but how had all that come about?

Cape Town, famous for its Table Mountain and harbour was first developed by the Dutch East India Company as a staging and supply post between Europe and Asia. Jan Van Riebeecks arrived in 1652 and started the first European settlement which later included German and Huguenot refugees. Of course the local inhabitants, the Xhosa weren't too happy about this and frontier wars began. During the Napoleonic Wars, Great Britain took over the Cape Colony and continued these wars. In 1820 British settlers arrived and claimed the land north and east thus engaging in conflicts with the African tribes of Xhosa, Zulu, Sotho and groups of Dutch farmers (Boers) who believed they had prior claims.

In the early 19th century the Zulu grew in power and expanded their territory under their leader, Shaka. The Materbele were an offshoot of the Zulu and they created a larger empire under King Mzilika, which included the Highveld (the area to the north-east). In the early 1800s the Dutch settlers, dissatisfied with the British rule in the Cape Colony, departed and started east on 'the Great Trek'. They migrated to what is now known as the areas of Natal, Orange Free State and Transvaal and founded the Boer Republic.

The discovery of diamonds in 1867 and gold in 1884 increased economic growth and immigration to South Africa. As one can imagine there was unrest between the English, Dutch and Zulus resulting in Zulu and Boer Wars. The Boers fiercely fought the British using guerrilla warfare and it was during the second Boer War that the English started the horrific idea of concentration camps when they imprisoned the women and children of the Boer fighters.

Eight years after the end of the 2nd Boer war an act of British Parliament granted nominal independence to South Africa and in 1931, with British power finally released, a United Party was formed seeking reconciliation between Afrikaans and English-speaking whites.

In 1948, with white people wanting more land, Apartheid was instituted by the National Party (mainly of Afrikaaners) and this became a legally ratified system of political and social separation of the races. The aim was to move all Africans to 'homelands' although they could work in the white areas of South Africa as 'guest workers'. Black Africans were not allowed to lease or purchase land, except in the reserves and this led to residential segregation. There was prohibition of mixed marriages which came under the Immorality Act. Non- white males were prohibited in certain areas of the country unless they were employed there.

By the 1970s when I was there, there was much opposition to this system and a number of people, both black and white were trying to change things. I was to see much evidence of this terrible segregation but, as a guest traveller, could not do anything myself but keep to the law of the land. I continued my journey to the economic capital South Africa, Johannesburg.

This town was born out of the Gold Rush in 1884. In the late 19th century it was a rough and disorganised place with miners of every nationality, European prostitutes and the local Africans doing any unskilled jobs, the Zulus taking on the laundry work. Of course beer was in high demand too. As the town grew and the value of land increased, tension developed between the British and the Boers, leading to the 2nd Boer war (1899-1902) which the British, despite attrition, finally won. An uneasy peace reigned in the city from then on.

Jo'burg is the world's largest city that isn't based on a river, lake or coastline. At an altitude of 1,753m high in the Witwatersrand, to the north and west are undulating hills in which the many springs, that attracted the Voortrekkers, drain water to the Limpopo and Orange rivers. It has a mild climate because of its elevation and in the summer, while I was there, there were afternoon thunderstorms and cool evenings. In winter, despite sunny days there can be frost at night.

INTO AFRICA – WITH A SMILE

I had an address to go to. The Welsh boys, John and Mike, were staying in a high rise flat in one of the suburbs with the two girls. Jane and Shilly had found work in a local coffee house, funnily enough the only one in Jo'burg to employ white waitresses, and the boys were also working, John in a hotel and Mike as an electrician. John and Shilly were now well and truly an item and did in fact later marry and live in Australia. I heard from them that Jacky had teamed up with Martin, another truck passenger, and they had worked in Jo'burg for a while too but had left to go back to Rhodesia where Martin had a teaching job. I didn't see Jacky again for quite some time.

I stayed a few days in the flat but the busy city with its tower blocks and freeways did not appeal to me. Also all the buildings in the white residential areas had bars on the windows and security gates and I could feel the social unrest. I did not venture anywhere near the black township of Soweto, originally built as a collection of settlements for black mine workers and now often raided by white Afrikaans police searching for 'troublemakers'.

Looking forward to finally reaching the ocean, I rode on towards Durban.

CHAPTER THIRTY TWO

Downhill to Durban

Coming down off the Highveld, the route that I took meant crossing the beautiful Drakensburg Mountains. This ridge snaked its way for several hundred kilometres from the North, inland, following the line of the coast, all the way through Natal and Transvaal to the Cape Province. En route I passed through Ladysmith, which, as every student of British history knows, was 'relieved' (the siege broken) during the Boer War. From Pietermaritzburg the road gradually descended down to the coast where the busy port of Durban lay on the Indian Ocean. What a treat to see it and realize that I was now on the pointy end of Africa having traversed it all the way from the blunt end at the top.

I had good friends to meet and stay with in Durban and, luckily for me, they rented a house in the posh, hilly suburb of Westville. Les Fleming and Keith Holmes were Saltbox MCC members who had also made the overland journey from the UK to South Africa but had done so, just a few years earlier, in a Landrover. They now had work in Durban and Keith's sister, Yvonne, had flown out to come and live there too. Les was in a relationship with a South African girl, Eddy, and there was another South African man, Peter, also sharing this big house. They welcomed me and said I could also stay but, although the house had four bedrooms, these were all taken so my sleeping space was under the dining table in the large living area. There I put my lilo and was quite happy to crawl into my sleeping bag there each night.

Les was a small, wiry cockney lad and he and his brother, Dave, had been very involved with the scrambling scene, competing on Dave's sidecar outfit in many events in southern England. Dave and I had been the best of friends before I left the UK for Australia in 1969. We had ridden together to many motorcycle rallies in Europe and he and Les always had a joke for every occasion. They were both

INTO AFRICA – WITH A SMILE

lively and well-loved members of our club. Unfortunately, a couple of years before, Dave, working as an electrical engineer, took a holiday to visit his brother, Les, in Durban. Les was well ensconced in SA life and was moving into this larger house. While the two brothers were shifting an electric stove, a terrible accident occurred and Dave was electrocuted, falling dead to the floor in front of Les' eyes. Shocked and grief-stricken, Les accompanied his brother's body on the flight back to England to be buried. Many, many people came to the funeral and it was there that I had re-met Les and told him of my intended ride down. He gave me his address and invited me to stay. So now I was set up with accommodation. Once again short of funds, I began to find my way around the area and look for a job.

Durban had more gentle feel to it than Jo'burg but had its own colourful history of scuffles between Zulus, Voortrekkers and the British. The land in the area had initially been granted by a Matabele chief to an English explorer, Henry Flynn, who had helped him recover from a stab wound but later the Boers and British, after fighting each other, had fought the Matabele and pushed them further north. Many settlers came from the Cape Colony and Europe and, when a sugar cane industry was set up, Indians were brought over to work and Natal has now the largest Indian population outside India.

In the 1970s many beachfront properties were built along 'the Golden Mile'. Durban was a holiday resort for the Jo'burg workers who found its subtropical climate a pleasant change and, of course, it had the added attraction of the sea. With warm, dry winters it was a convenient place to have a holiday home there to take break from chilly Jo'burg. There was a pleasure-craft marina as well as the important international port that had grown from trade with the commercial capital.

An article in the local paper about my journey through Africa gave me some kudos and I approached the local motorcycle dealer, Charlie Young, to ask for a job. They didn't really have a vacancy but took me on as an aid to the sales team. I could help chatting to customers and delivering the bikes. An unusual ride I remember was taking one of the little fold-up bikes to a yacht owner.

While working at the motorcycle shop and meeting all kinds of people I became aware of the term used for most women, 'Aunties'. I suppose it was similar to the Australian expression 'Sheilas' but it seemed to have a different connotation. 'Aunties', the South African white women, put on a kind of pedestal by their men, were supposed to look and act like ladies and were somewhat revered, NOT like 'Sheilas'. In accordance with this attitude, the women tended to be careful of their dress and make-up and mostly behaved in a lady-like fashion. Of course this was almost impossible for me but I did my best and with money coming in from wages, bought a few new T-shirts and another dress for work.

One day, early in my new career in the shop, a funny-looking character came in and said that he had seen the newspaper article and wanted to meet me. He was about 5'10 in height, of slim but well-muscled build and had very fair skin and receding hair. Dressed in dirty white overalls, he had twinkling blue eyes and a huge grin. His broad Afrikaans accent and ready laugh were irresistible. His name was Gene Visser, and he too owned an old BMW but it looked nothing like mine. It had a purple metal-flake tank, chromed engine parts and mud-guards, ape-hanger handlebars and exhaust pipes that came straight out and up above the head level. He invited me to come and meet the local bike crowd on the next breakfast run.

" Yesses, Linda, come for a lekker jawl with the okkies on Sunday." All I understood was breakfast run. Breakfast run? What was all that about? I'd never heard of such a thing. Thus began my introduction to the Durban bike scene.

CHAPTER THIRTY THREE

Durban Delights

It appeared that on Sunday mornings the motorcycle fraternity met outside Honda House in Pine Street where they would pose for a while with their gleaming (or otherwise , in my case) machines. Among them I spotted at least two BMW R69Ss, an R50, a Norton Commando, several Triumphs and other, Japanese, machines. There was another young woman, who they called Flame Lily (she had red hair) on an R50 too; all in all about 30 bikes, depending on the weather. Due notice was taken of paint jobs, accessories and innovations; I was amazed to see some-one showing off a big top-box with a television fitted inside. Quite why was beyond me but he seemed proud of it.

At a given signal the machines would file out and follow the leader on the road out of town, beginning the ascent to the destination of the Rob Roy Hotel, about 85miles away in the Valley of a Thousand Hills. Some rebels would overtake and show off but they eventually all came together at the restaurant to tuck into fried breakfast, breaking into smaller groups to chat. What amazed me was that after breakfast most of the crowd just went home. Many of the riders were solo; they had left their wives at home with the kids and would be re-joining them for lunch. Back in the UK or Australia we went out on club runs for a whole day together, having breakfasted before we left home, so therefore this idea was new to me.(Now it is normal in the Australian bike scene) However, once I'd got to know some of the riders I did indeed share some longer rides with them.

Gene used to drop in regularly for a chat at the showroom where I worked. He invited me home to meet his very large family; he had thirteen siblings and I was amazed that his mother was still alive! An Afrikaans family, they lived in a small house at the Point, a suburb close to the docks where he worked as an engineer on the tug and pilot boats using his considerable mechanical skills. Although I am allergic to ships in general because of their motion, I was fascinated to be taken on board these working boats and see the big diesel engines that Gene kept in such good order. He was well respected amongst his colleagues; apparently at one time he had taken a huge risk when an engine had been on the point of explosion and his quick action had saved both the boat and the crew. His relationship with the black Africans who worked under him was also good as he often helped them fix their small motorbikes and laughed and joked with everyone.

One evening, I braced myself to go on board the pilot boat with Gene and his crew as they took a trip out into the bay to visit some of the big container ships that were waiting their turn to dock and unload. Sometimes they were moored out for days or even weeks and the pilot boat would take out their mail for them. Our little boat pulled up alongside these huge vessels and a ladder was put over the side down which a sailor came with a bulging bag which he exchanged for the one that Gene handed up to him. The crew of the container ships were so happy to receive their mail that they would give a token of appreciation. After visiting several ships Gene and his men would share out the booty of several different types of liquor i.e. vodka from Russian or Polish ships, brandy from Spanish, etc. Good for settling my stomach! This postman's job wasn't always an easy exercise; sometimes the seas were high and it was quite dangerous to pull up alongside these heaving monsters to make the exchange.

One day there was a huge event at the harbour; The QE2 was coming in. Gene was to go out on one of the tug boats to pull her in. I asked if I could come too but his answer was. "Not a chance, knowing you, you will be throwing up over the side and everyone will see!" He was right to be worried about the audience; it was literally thousands of people who came to the harbour to watch the big ship come in. My seasickness wouldn't have added to the ambience.

On one of the many ships that called into port there was an American, a ship's mate named Harry and he belong to a group in USA called the Retreads. This is the equivalent of Ulysses in Australia i.e. it is a club that is for riders over 40. He was from California and owned a big bike there. Harry had contacted the Durban bike fraternity and asked if he could come with us on one of the breakfast runs. It was arranged that the group would pick him up from his ship. I am not sure why but somehow I was elected to take him

INTO AFRICA – WITH A SMILE

pillion on my BMW. This was not a good idea. In his fifties, he was a big man and although I didn't mind his presence the bike did and really struggled up the mountain roads. When we arrived at the breakfast destination, I insisted that he be transferred to a more appropriate steed. This done it was a much more satisfactory ride for me and we were all rewarded when Harry, with the usual American bonhomie, invited us back on board his ship for evening drinks.

CHAPTER THIRTY FOUR

Life in Westville

Meanwhile, in Westville, where I was staying, I was learning to be an 'Auntie'. The domestic arrangements were somewhat different from other share houses in which I had lived. The main difference was that we didn't have to argue about the housework- we had a maid. A young Bantu girl, her name was Precious and she lived in a little hut in our backyard. Precious started the day by bringing us all a cup of tea in bed – yes, I was served mine while lying in my sleeping bag under the table. "Tea, Madam" were the first words I heard every morning. She then prepared our breakfasts, if we so wished, and busied herself during the day cleaning the house, washing our clothes and preparing the evening meal. Then she retired to her own quarters. I was quite taken aback by this situation. Of course the same thing had applied in my friend's houses in Rhodesia with their housemaid but here we were all English! (apart from Peter and Eddy).

I asked Yvonne, who had now been living here for about a year, "How can you do this, Yvonne, have a maid, for goodness sake? We can do our own housework."

"But, Linda, we had to, she needs the work".

It seemed that if you didn't have a maid everyone would know and there would be a constant stream of enquiry from the native girls who were desperate for a position. The African population were not allowed to live in Durban unless they had work there. If they were earning a wage they could save money and, when allowed by their employers, go back for a trip to their homelands and take money and goods to their families. Some women even left their young children with their parents to look after while they worked in the city. It was a cruel part of the Apartheid system and would not be changed for many years after I was there.

Precious was now in employment with us and could proudly tell

INTO AFRICA – WITH A SMILE

all other applicants to go away. She had her own space and we provided food for her. Her main diet was mealy-meal (corn) and we bought a sack of it for her whenever she requested. The household had employed several maids before her, some more reliable and less light- fingered than others. Yvonne advised me to keep any valuables well hidden. She was sympathetic to the maids' needs and tried to alleviate their trying circumstances in any way she could. One girl she had to rush to hospital when she discovered her screaming and bleeding almost to death in her hut after an unsuccessful abortion.

The Apartheid system was hard for us to swallow, used as we were to the mixing of races in the UK. Here the classifications were White: Europeans and white South Africans, Black: native Africans and Coloured: Asians and people from mixed race. This was a general guideline and many people who weren't sure of their classification lived in fear of being punished under the Immorality Act. Some had left South Africa rather than be separated from other members of their family. Of course I was unhappy with these ideas and realised what terrible injustices were being applied but, as I was not a permanent resident and had no voting rights, I had to accept the status quo and just enjoy having my housework and washing done.

We all seemed to get along well. Everyone in the household was working and so we were out during the day and congregated around the dining table to chat over our evening meal. Sometimes Peter would come in late in the evening after a drinking session with his mates and, if I was still awake, under the table, we would have a chat before he retired. His bedroom door was just opposite my bed-site.

Westville was an upmarket suburb and our street was full of large houses with well-maintained gardens (by the houseboy, of course) and good views of the city below. One day, as I was turning into the drive on my bike an 'Auntie' hailed me. She was a middle-aged, well-dressed lady who lived a few doors away. She said that her husband collected old vehicles and would be pleased to meet me. In the meantime would I care to join her today for afternoon tea? As it happened it was my 29th birthday and I was delighted to accept and have an unexpected treat. As we sat on her shaded veranda by her swimming pool drinking tea from bone china cups and eating cake and cucumber sandwiches – served by her housemaid, of course- she said that her husband was shortly hosting a meeting of the Vintage and Veteran Car and Bike club and I would be welcome to attend.

153

LINDA BOOTHERSTONE

When the day came I was amazed at the collection of well-restored vehicles that appeared. The owners and their wives (the Aunties) were all very well-dressed and we sat around drinking South African wine as the day mellowed into a balmy evening with the lights of Durban twinkling below. I was also impressed by the candles in their lily-shaped holders that floated in the pool. It was something else that I hadn't seen before and I thought them very sophisticated.

Another neighbour of mine in this suburb was Pauline Hailwood, wife of the famous motorcycle champion, Mike the Bike, who I had met at the BMW promotion in London. She and the children were in South Africa at the time and the newspaperman who interviewed me had put me in touch with her. After phoning, I went to visit Pauline for afternoon tea and she told me that Mike was still in Europe. He had recently had an accident while racing cars at the Nurburgring in Germany and was recovering.

Pauline, Mike and their child, two year old Michelle, had not been in Durban long. They had rented this house out while they were away from the country and, apparently, the tenants had not left it in a very good state so Pauline was in the process of refurnishing. She was a very attractive brunette, in her mid-thirties and had been a professional model, an air hostess and had even appeared in some TV programs before she lived with Mike. Now having a young child to look after, she said it was a big effort to keep up the 'model look' all the time and preferred to be more casual while she was at home. Of course, when in the public eye with Mike she had to dress up. Fortunately, I didn't have this pressure on me. She was very interested in my trip and I spent a few pleasant hours with this warm and down-to-earth woman. Unfortunately, just a few years later she suffered a terrible tragedy.

While living with Mike and the two children in Warwickshire, UK, they ordered fish and chips from a shop in the neighbouring village. Mike and the kids jumped in his Rover RDI to go and collect them and on the journey through Portway a rubbish truck illegally crossed the central reservation and his car hit it. Their little girl, Michelle, aged 9, was killed instantly and Mike, mortally wounded, died in hospital two days later. Their son, David, aged 5, was injured but fortunately survived. The truck driver was fined £100!

The motor cycle world was shocked and distraught and, up until 2011, there were Mike Hailwood memorial runs in the UK every year. An odd thing was that, apparently, a fortune teller in South Africa had told Mike that he would die before age 40, killed by a truck. He had just turned 40 in March 1981 when the accident occurred. Pauline was fortunately supported by Mike's family and much later ran a business in the Andalucian Spanish hills not far from where I later lived. She and her son continue to be involved with motorcycle racing helping with various promotions.

Chapter Thirty Five

More Durban Delights

Although we shared many common interests, the rest of my household were not involved in motorcycling and I spent most of my weekends with Gene and the Durban Motorcycle Social Club. One of its members was a keen historian and he was interested in taking us on a trip to Spieonkop, an important battle site during the fight for the relief of Ladysmith in 1900 during the 2nd Boer War. He promised that he would fill us in on all the gory details when we arrived. It was about 100 miles away and entailed a very pleasant ride up through the foothills of the Drakensburg Mountains. The last few miles were on a steep, uphill, stony, dirt track and I fell off a couple of times before we reached the site of the battle and were given our history lesson.

In January 1900 the two generals in charge of the troops were General Redvers Buller for the British and Lois Botha for the Boers. Spieonkop, at 1,410ft, lay in the centre of the Boer lines and was an important strategic position. If the British captured it they could, hopefully, break the Boer hold on Ladysmith which was under seige. With a troop of Scottish and English soldiers they made a surprise attack in the early morning. They climbed the incline through the mist and found a smaller Boer force which they easily overcame and killed. They then dug in their positions but, to their horror and ultimate demise, when the mist lifted they realised that they had not reached the summit as there were other 'kops' (heads) of the range above them. The Boers, positioned there, were able to fire down on the British position, taking aim easily and the men below were having to use the bodies of their fallen companions as barricades. Throughout the skirmish for control of this and other positions during the campaign there was one mistake after another in the instructions of both the British and the Boer generals. In this particular battle the Boers came off best but the British finally won and broke the siege of Ladysmith with a larger force four weeks later.

INTO AFRICA – WITH A SMILE

The siege had lasted 118 days.

We wandered around the area of the battle, the small hilltop overlooking a grassy plain, where the first troop had hopefully climbed and then sat, with the Boer bullets raining down on them as the dawn broke and the mist cleared. There was no shelter or escape.

As many of the men had been Scottish, I sang 'Nut Brown Maiden', originally a bagpipe tune that the regiments often played. It was my small offering in their memory, for we all felt the spirit of the men around us and thought of the terrible wastefulness of war. Two ultimately famous men who were involved in this battle were Winston Churchill, who was a courier and Mahatma Ghandi, then a stretcher bearer.

Another less dramatic ride I took with the motorcycle club was to the races at the Roy Hesketh circuit in Petermaritzburg. It was a fine day and only a short ride. The atmosphere at the track was similar to that of the circuits I had attended in Australia; we could roam around the pits and chat to the riders. It was a fine day and the racing was good, but I cannot recall the names of any of the riders of the time.

A longer journey that I took over a weekend with Gene was out to another part of the Drakensburg range to a place near the tourist resort of the Drakensberg Gardens Hotel. It was in the National Park area about 200 miles from Durban so we took our tents to camp overnight. The last few miles were on dirt road and, due to recent rains, they were muddy. My front mudguard jammed with the mud and I slid over. Unharmed, I tried to pick up the bike but needed the help of Gene who was busy laughing and taking photographs of my struggles. Typical.

At one stage of our exploration, while walking around the area, Gene spied a night adder and decided he would catch it as it could be sold to the hospital for its venom to be used to make anti-venom for snake bites. He tried to corner it and the snake actually jumped up, out of the grass, about two feet in the air to attack. I was already about twenty feet away, running! Much to my relief he didn't catch it.

Another event that I rode to in company with the club was the Buffalo Rally held in March, near Bloemfontein in the Orange Free State. It was the largest and most well- known rally in South Africa and though I had expected a large crowd, was overwhelmed to see about 3,000 motorcyclists camped over a wide area. There were all different types of riders and machines but the most worrying thing was that many of the boys had guns and were having shooting contests; setting up empty beer bottles and firing at them. Of course they had emptied the bottles of beer – down their throats -so their aim was not too good and this was amongst the tents! I heard of at least one incident involving gun-shot wounds and I think the first-aid tent was also kept busy with other injuries from fist fights. The local pubs complained of their premises being damaged but, overall, the town made a great deal of money over the rally weekend so the police mainly turned a blind eye.

It certainly was an eye opener for me though. At that time most of the rallies I had attended anywhere in the world had not attracted more than a few hundred participants (apart from the Elephant Rally in Germany) so this was amazing. However, I entered into the swing of things and had a few beers myself. There may have been awards given for long-distance, best bike etc. but I don't remember them! With my lagerphone I joined in with a few sing-songs.

My repertoire of Afrikaans songs had been growing and included 'Me Sarai Marai', 'Me hartje , me liefe' and 'Bobbejaarn clem de burg'. These I had learnt while attending a folk club that I had found at Umlunga Rocks, a suburb in the north of Durban. The participants were the usual folky types; friendly and keen to share their interest in songs and stories.

The most well-known folk singer/songwriter at that time was Jeremy Taylor. He came to South Africa from the UK to work as a teacher and began singing in clubs in Jo'burg in the early 1960s. He had a huge success with his song 'Ach, Pleez Daddy' a light hearted ditty using the Afrikaans accent and which was inspired by the expressions used by the children he taught in the Jo'burg suburbs. His recording of this song sold more copies in South Africa than any of the current Elvis hits. Unfortunately the South African government did not take kindly to his mix of Afrikaans and English (which many people used in their everyday life) as they felt that their language should be kept 'pure'. After returning to the UK in 1964 he found that he was not allowed to re-enter South Africa and was not admitted back until 1979. So, unfortunately, he was not in the country while I was there but many folkies were singing his songs and I soon learnt the words. To this day 'Ach, Pleez Daddy' brings a smile to any South African's face. Jeremy Taylor returned to live in the UK in 1994 and is still touring on the folk music circuit.

Into Africa – With a Smile

CHAPTER THIRTY SIX

A Make-over for the Bike

I was really enjoying my time in Durban. I had a comfortable place to stay, a job and good friends. In the sunny, sub-tropical climate nearly every day was fine. I was there in the rainy season but the rain only came in short, sharp showers in the afternoon, not like the constant drizzle in the UK and I never had to wear heavy wet-weather gear. The Durban climate did have one drawback though, it was humid and the air was salty due to its proximity to the sea. This was playing havoc with my bike which was getting more rusty and corroded by the day.

"Yessus, Linda, it's a mess. We need to fix it, ek se" was Gene's opinion. He had taken a shine to my BMW and was offering to help me rebuild it. I immediately accepted. So we took the bike to his place at the Point and the strip-down commenced. The engine was taken out of the frame which was re-enamelled black. The many dents in the tank, from when I had dropped it, were filled and it, together with the mudguards, were sprayed with a dark green metallic paint. Gene wanted it done in metal-flake but I said "No!" I also refused to have the tappet covers chromed; they were, along with the barrels and heads, sandblasted. The only chroming I allowed was the air-filter cover and the headlight rim. The wheels were re-spoked and the final classy touch was a gold pinstripe line on the tank and mudguards and a map of Africa, also in gold, on the tank sides. Unfortunately I could not get new pipes and silencers and these were replaced later in London by BMW.

During the time my bike was off the road, Gene sometimes lent me his BMW. On these occasions I hoped no-one I knew would see me. With the horrible ape-hanger handlebars, that I could hardly reach, the bright purple paintwork and the high straight-through exhaust pipes I felt very conspicuous and a right idiot! However, he also lent me his bakkie (ute).

INTO AFRICA – WITH A SMILE

The town of Durban was just beginning to develop in the early 1970s; a few tall towers and the beginning of a freeway – the traffic was light and it was easy to find my way around.

Despite the obvious signs on buildings and beaches ; 'Whites only' or 'Non whites only' and segregation in buses, the town seemed to be quite peaceful with no obvious signs of crime or discontent and the violence that was coming. There was one incident, however, which made me feel very uncomfortable.

One day, while driving Gene's bakkie, I clipped an African cyclist who wobbled out in front of me. He fell off and, although appearing unhurt, was picking up his bike. I stopped and jumped out of the car, rushing over to help him but immediately an Afrikaans policeman came on the scene, grabbed hold of the man and started abusing him for getting in my way.

"So sorry, madam," the policeman said, "Do you want us to charge him?"

I was appalled, he was about to march this poor man off to the police station.

"No, no, not at all, it was my fault! Please let him go."

The policeman reluctantly released him and the man hurriedly left, pushing his bike. I felt really embarrassed and horrified that, because I was white I could do no wrong.

Apart from the songs I had learnt, my friendship with Gene and his family and biker friends had taught me many of the South African slang words and colloquialisms which I soon began using myself.

Graze - I had learnt this already in Rhodesia and was a term used for having a snack or even a meal.

Kudu bars – Protective bars on the front of vehicles, like 'roo' bars in Australia.

Takkies – sneakers

Cokie pens – felt tip pens

Bakkie – utility vehicle – 'ute' in Australia

Jawl – a good time

Lekker – good

Biltong – dried meat like jerky.

Sonnerlikkie – sunglasses.

There was also a popular song amongst the bikers which was a version of *Heaven is my Woman's Love*.

It went; Heaven is my motorbike, heaven is my motorbike

161

It gives me cause to jawl all day, in licence fine not much to pay....

I began to hear myself slipping into the Afrikaans accent, a distinctive way of speaking. The language had come from the Dutch but had developed differently. The South Africans descended from the 1820 British settlers did not have such a strong accent but still pronounced their vowels differently, somewhat similar to the New Zealand manner of speech.

Finally my bike was all together again and it looked absolutely gorgeous. Gene had also checked the engine so it not only looked but performed like a new BMW and was quite the most distinctive one around. My employment at the motorcycle shop came to an end. They had made a job for me but my presence had not made much difference to their bike sales so they had decided to 'let me go'. As the BMW was now in tip-top condition and I had spent about six months in Durban, I felt it was time to move on and complete my journey. A strong incentive was that I had heard that the famous Isle of Man Tourist Trophy races (T.T.) was going to be terminated after this year (1975) so, as I had never yet attended, I felt I should be back in Britain to see them. I therefore booked a ship on the Union Castle Line, leaving May 6th from Cape Town. If I left Durban just after Easter it would allow me time to ride along the coast to include 'the Garden Route' to the final city on my African adventure.

"You rrrrat, Linda" were some of the final words that Gene had to say to me as I heavily loaded (where do these extra possessions come from?) my bike and rather sadly said goodbye to him, his family, Les and Eddy and all the other good friends I had met in Durban. It is always with mixed feelings that one leaves a settled life to be on the road again. Blinking back a few tears I turned the BMW West towards the Cape Province.

Into Africa – With a Smile

Durban harbour

Durban bikers run

Rob Roy Hotel, Durban run

A ride with Gene and Hennie

Workshop in Durban

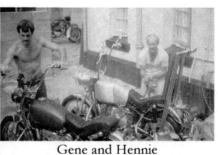
Gene and Hennie

Oops

Gene

Linda at Drakensburg Mountains

Sign for Spieonkop

The ride up to Spieonkop

Shooting demonstration at Spieonkop

Falling on the road to Spieonkop

View from Spieonkop

Warning!

Gene fixing Linda's bike

Linda

Queen Mary in Durban harbour

Linda with her rebuilt bike

Mike Hailwood and Pauline with baby Michelle

CHAPTER THIRTY SEVEN

The Coastal Route

As usual when going on a journey to a new place, other people are quick to give advice on it, even if they haven't had personal experience. The route along the coast would take me through the Transkei, one of the two homelands designated for the Xhosa people. It was given nominal autonomy in 1963, although this was only recognised by South Africa; the UN called it 'sham independence' and 'invalid'. Most of the population were recruited to work in the South African mines. At that time it had a Prime Minister, Chief Kaiser Daliwonga Matanzima, who was a nephew of Nelson Mandela.

The comments I received before I left were "Be careful going through that black's country, it's very dangerous." Of course I ignored such an attitude and was looking forward to my new adventure.

The route was magnificent; following the view of the sparkling blue waters through rolling hills leading down to beaches in between rocky headlands. The countryside was dotted with brown and white rondavels with thatched roofs and cattle roamed on the foreshore and beaches. The local women were the busy ones, as usual, walking in their colourful apparel with loads carried on their heads. They flashed white toothed grins as I rode by.

My first exciting experience was when I came to the beautiful estuary of the Umzimruba River at Port St John. On either side of the river mouth were spectacular sandstone cliffs and there was a road along each bank. While riding on the eastern side I saw a large construction on the bank which had signs saying, 'Keep out, film personnel only. Property of Shepparton Studios.' As I was riding slowly by, craning my head to try and see what it was all about, (my granny had lived in Shepparton and I was aware of the film company there) there came a shout from above. I looked up and two men were

INTO AFRICA – WITH A SMILE

waving at me.

"Oi, where are you from?" they had seen the GB plates on my bike.

"From Surrey, "I replied.

"Come on up."

"But it says 'no entry'."

"Naw, that's for the South Africans, you're English, you can come aboard."

I parked the bike and carefully found my way up the scaffolding and came to be standing on board a big battleship – made of hardboard and wire! The British workmen were building a mock-up of the 'Blucher', a German ship, for the film of Wilbur Smith's book *Shout at the Devil*. In this scene it is undergoing repairs while moored in a river mouth, supposedly in East Africa. The construction was truly amazing. With hardboard, wire, paint and rivets they had made it look substantial – from a distance. They told me to ride along the road on the other bank to see the full effect which, after thanking them profusely, I did. It really did look the part. The crew had advised me that Roger Moore and Lee Marvin were involved in shooting another scene the next day at a site further inland so I went along to watch. I would love to have been involved as an extra but I wasn't the right colour to be an Askari warrior, the ones needed for that particular part of the movie. It was thrilling to be so close to the stars even though they were too busy for a chat.

I had a WIMA contact in East London, a port along the coast. Her name was Deirdre and she had invited me to stay. She was a small, dark- haired, skinny girl who proudly owned a huge red 750 Ducati. Owing to her slight build, to start it she had to take an enormous leap to put all her weight on the left kick start lever. I watched in amazement every time she did so. Her grin of achievement when the engine snorted into life was a joy to behold. She and her university friends invited me to a party where a good deal of 'Malawi Grass' was handed round. I did try some but it just gave me a sore throat and I didn't reach the giggling stage that affected the others. There appeared to be quite a culture of 'dagga' smoking amongst the students but it was something that had never appealed to me. Thanking Deirdre for her hospitality I continued a little further along the coast.

My next stop was to visit another WIMA member, the South

African captain, Lil Collet, who lived in Port Elizabeth. Lil and her husband were originally from Scarborough, UK and were in their 40s. They hosted me for a couple of days, taking me around the previous British colonial town which had many fine, old buildings and a statue of a horse commemorating the tens of thousands of horses and mules that died during the terrible conflict of the Boer Wars. Port Elizabeth was also the site of the concentration camps that the British set up to intern the captured Boer women and children. (a horrible reminder of the inhumanity of war.) From Port Elizabeth I continued on through Knysna and then turned inland at George to visit Oudtshoorn, 'The ostrich capital of the World'.

The town was started in 1858 but a lack of water made the land unsuitable for cropping so a local farmer started ostrich farming. Ostrich feathers had become fashionable amongst European nobility and there were feather auctions conducted at nearby Mossel Bay. Between 1875- 1880 the price of ostrich feathers soared – fetching up to £1000 a pair! They were worth, in weight, almost as much as diamonds. Farmers planted lucerne as feed for the birds and the feathers were dubbed 'White Gold'. Over- production led to a slump and in 1885 there was bad flooding in the area and the town was in serious depression. However, things improved and another boom in feathers occurred after the second Boer War. But, again, the market collapsed in 1914. Apart from the effect of WW1, the advent of open-topped cars made the wearing of ostrich feathers impractical. 80% of ostrich farmers went bankrupt and the birds were either slaughtered for biltong or set loose to fend for themselves. I was keen to see these birds and, at that time, it was possible to climb aboard and go for a ride on one, which I did. Quite different from riding the BMW and the closest I would get to wearing ostrich feathers.

I left the tickly attractions of Oudtshoorn to take a trip out to the Cango Caves- limestone show caves which are part of the huge cave complex in this area. I continued along the 'Garden Route' from Mossel Bay in the Western Cape (little Karoo.) This area is so named because of its rich, green, ecologically diverse vegetation. It has a very mild climate with rain year round because of its proximity to the Outenga and Tsitikama mountains, just inland from the coast. Their run-off feeds the myriad of plants along the coastline. It was very pretty and I thoroughly enjoyed the ride into Cape Town.

Chapter Thirty Eight

Cape Town

Well, here I was at last. There was no mistaking the place; here looming over everything was the famous Table Mountain, its end peaks of Devil's Peak and Lion Head forming bookends for its level central plateau. It was capped, as usual, with its 'table-cloth' of cloud. What a spectacular sight! I had last seen it when my ship, the 'Oriana', taking me to Australia in 1969, had docked in the huge harbour. Now I had reached the city under my own steam, or rather a 500cc petrol engine, and I was quite overcome with pride and an enormous sense of achievement.

I was just approaching the main part of town when another BMW rider drew up alongside me at the traffic lights.

"Wow! have you ridden that from England?"

"Yes."

"You must come home and meet my wife."

I was then escorted to Erik and Hannie's lovely modern house in the Cape Town suburbs and was told that I would be staying with them.

Erik was in his thirties and a wannabe sailor. He had put down the keel of his boat in the back yard several years before but, still working as an accountant, had not got much further with the boat building. However, the house was littered with such items as compasses, charts and a ships clock. Many paintings and photos of ships adorned the walls and books on sailing techniques filled the book shelves. Erik and Hannie were very kind and good company, feeding me and showing me around town. The only characteristic that I found disconcerting was that they were born-again Christians and tried very hard to lead me to the Lord. A lost cause, I'm afraid.

The first thing I wanted to do was find the offices of the Union Castle Line, check on my booking on the Pendennis Castle and see what I had to do to get myself and the bike aboard when the time came.

It was at these offices that I met Devon Dold and Bruce Stephens and I must say initially I wasn't impressed. They were both wearing black leather jackets and the 'colours' of their local motorcycle club, the One Percenters. These consisted of a sleeveless blue denim waistcoat with a picture of a skull, with a bike crashing through it, painted on the back. This apparently was a copy of the logo of the old Johannesburg Hell's Angels from the days before the Angels had world-wide copy-right on their designs. Although dressed in these, for me, off-putting bikie clothes, Devon was very good-looking. He was tall with dark, wavy hair and spoke with a posh English-South African accent. Bruce was shorter, with sandy-coloured hair and didn't speak so well.

It transpired that they too were taking the ship to the UK with their motorbikes as they also wanted to see the TT races and afterward tour the British Isles before finding work there. Word of my arrival in Cape Town had been given to them by Dierdre from East London who had lately ridden there on her Ducati for a holiday. While she was in town I went on a few rides with her, taking Hannie along while Erik was out work.

After we had checked the procedure at the shipping office, Bruce and Devon invited me to their flat for coffee and introduced me to their flat mate, Dave Cummings. The first sight I had of this person was in the kitchen where he was peeling onions wearing a diving mask, explaining that it stopped his eyes from watering. Apparently he was the designated cook and made several pies at one time to feed the others in the share house. I was told that the food kitty at that time was a mere one Rand per week each, so I am not sure what the ingredients were!

We had been told that we didn't need crates made for the bikes, that they would just be craned into the hold and, hopefully, well tied down amidst the general baggage. Since there was little else to arrange, I spent the next few days having a look around town and tracking down some more WIMA members.

I found a small but enthusiastic group of women in the Cape Town section, ably led by Claudia Huenis and her boyfriend, Ron Hines, who also had BMWs. We met in town one day to go out on a ride to Stellenbosch. One of the best known towns in the Cape Province, it is celebrated around the world for its wines. Its founder,

INTO AFRICA – WITH A SMILE

Simon van der Stel planted the area with oaks which are now magnificent old trees. Huguenot refugees settled there in the early 1700s and planted grape vines in the fertile valley of the Eerste River. The wine industry boomed, gracious Cape Dutch houses were built and a wine route was established in 1971. We made good use of it!

One day at Poste Restante I collected a letter which brought a few tears to my eyes. I had left my husband, Terry Bick, back in Perth, Western Australia in 1972. From there I originally returned to the UK because of the serious illness of my father. While in England, I decided that my marriage to Terry was a mistake and that it was better that I made a clean break and not return to Australia. He was a good man, with his own business, but he had married the wrong woman. My feet were far too itchy to allow me to settle down and I should have known that and not made a promise I couldn't keep. In this letter he was asking for a divorce; he had found someone far more suitable and wanted to re-marry. Would I sign the enclosed papers? The divorce would be on the grounds of desertion. Well, of course I would sign but the request in the letter stirred up all sorts of emotions, especially guilt, but also the memories of the good times we had together in Perth and what a really decent person I had left. Well, I had my little cry, then signed and posted the papers. Another part of my life was well and truly over and now I must look forward. Drying my eyes I carried on exploring the Cape environs finding out as much as I could before my departure.

CHAPTER THIRTY NINE

Father of the Nation

Of course I was only too aware of the international disapproval of the South African Government and its apartheid system. Also that there were various groups fighting against it but I had not closely followed the people involved. Strangely enough, although he is now world-renowned, the name Nelson Mandela was not, in 1975, on everyone's lips. I was not informed, while I was in Cape Town, that he was imprisoned on Robben Island, just a few miles offshore, or of his continuing struggle for justice for his people.

Nelson Mandela was a Xhosa, born to the Thembu Royal family in 1918. He was well educated, studying law at university in South Africa and, while living in Johannesburg, became involved in anti-colonial politics, joining the African National Congress (ANC). After the Afrikaans National Party established Apartheid in 1948, he rose to prominence in the ANC and was repeatedly arrested for seditious activities. Although initially committed to non-violent protest, he co-founded a militant group, Umkhonto we Sizwe (MK) and led a sabotage campaign against the Apartheid government. In 1962 he was arrested and sentenced to life imprisonment.

From 1964-1982 he was kept on Robben Island where he and other prisoners spent nights in small, damp cells, 8'x7', and days breaking rocks into gravel and later working in a lime quarry. Not allowed to wear sun-glasses, the glare from the lime permanently damaged Mandela's eyesight but he struggled on in the evenings studying for his higher law degrees. Initially classed as the lowest-grade prisoner he was permitted one visit and letter every six months and, although thus isolated from the outside world, he read smuggled news-clippings. He and his fellow political prisoners held debates looking at all aspects of politics and religion, initiating the 'University of Robben Island' whereby prisoners lectured in their own areas of expertise. Mandela also studied Afrikaans hoping to build mutual

INTO AFRICA – WITH A SMILE

respect with his warders and convert them to his cause. As time went on various official visitors met with Mandela, including representatives of the South African Progressive Party and the British Labour Party MP Dennis Healey.

Mandela was not allowed out to attend the funeral of his mother or of his son, who was killed in a car accident, and first saw his daughters after ten years' imprisonment. Early in the 1970s Mandela co-operated with a new prison officer to improve living conditions and the inmates were allowed to play some games, such as football, which lifted their spirits. By 1975 Mandela's prison status had become higher and he was allowed more visitors and letters and he corresponded with anti-apartheid activists such as Mangosuthu Buthelez and Desmond Tutu.

The world focus on the riots in Soweto in 1976 brought the young political activist, Steve Biko onto the scene. A founder of the South African Student Association he was also working for the acceptance of black people on an equal footing. He coined the phrase 'Black is beautiful'. In August 1977 he was arrested on a charge of terrorism, tortured and finally beaten to death in a jail in Pretoria. Because of his high profile, the news of the manner of his death spread quickly and a liberal journalist, David Woods, published a book about his life which was made into the film, *Cry Freedom*.

Following the international outcry over Biko's death, interest was renewed in Nelson Mandela's plight. In 1978 when he celebrated his 60th birthday he began receiving awards from various countries. In March 1980, the *Free Mandela* slogan was developed by journalist Percy Qoboza, sparking an international campaign that led to the UN Security Council calling for his release. The S.A. Government refused; they relied on powerful cold war allies, US President Ronald Regan and UK Prime Minister Margaret Thatcher, both of whom considered Mandela's ANC a terrorist organisation sympathetic to communism and therefore supported its suppression. Finally, international pressure led to Mandela's release in 1990 and, amid escalating civil strife, Mandela joined negotiations with South Africa's Nationalist President F W De Klerk to abolish Apartheid and establish multinational elections. In 1994 he led the ANC to victory and became the first black President of this country.

Mandela spent the rest of his political career improving life for all in S.A. introducing a new constitution to encourage law reform,

LINDA BOOTHERSTONE

combat poverty and expand health care services. He was well respected overseas, often acting as a mediator in international disputes. He declined a second term in office and became an elder statesman focussing on charitable works and instigated the Nelson Mandela Foundation to combat poverty and HIV/AIDS.

Mandela was a controversial figure for much of his life. Denounced as a communist terrorist by critics, he nevertheless gained international acclaim for his activism, receiving many honours including the 1993 Nobel Peace Prize. He is held in deep respect in S.A. and often referred to as 'Tata' or Father of the Nation. His death in 2013 was mourned by millions both in his home country and throughout the world and his incredible fortitude and forgiveness of his oppressors will always be remembered. His autobiography, *Long Walk to Freedom*, was first published in 1995.

CHAPTER FORTY

Matters of the Heart

One event that had received a great deal of publicity around the world a few years earlier was the first successful heart transplant performed, in Cape Town, by Dr Christiaan Barnard. So, I was aware of this celebrity and duly rode around to see the Groote Schuur hospital as well as the ivy-clad buildings of the Cape Town University.

Christiaan Barnard came from an Afrikaans family and his father was a minister in the Dutch Reformed Church. Interested in medicine from an early age, he became an intern and served his residency at the Groote Schuur hospital. Proving to be a skilled surgeon he made his first successful kidney transplant in 1953 in the USA and, after experimenting with animals, performed the first successful human heart transplant in December 1967 assisted by his brother, Dr Matthias Barnard. The operation lasted nine hours and used a team of thirty people. Although the patient died of pneumonia after eighteen days the operation had passed a milestone in a new field of life-extending surgery.

Barnard was celebrated around the world for his accomplishment. He was good-looking, debonair and photogenic and enjoyed the media attention following the operation. International fame took its toll on his personal life; he became quite a playboy, married three times and had an extra marital affair with film star Gina Lollobrigida. From his marriages he produced six children.

Barnard performed many cardiac transplant operations with varying degrees of success, trying different methods and improved immune-suppressant drugs. He retired as head of Cardiothoral Surgery in Cape Town in 1983 after developing rheumatoid arthritis which ended his surgical career. He had by this time become involved in anti-aging research and spent time as a research advisor in a clinic in Switzerland where the controversial 'rejuvenation therapy' was practised. He spent his remaining years living both in Austria, where he established the Christiaan Barnard Foundation dedicated to helping underprivileged children throughout the world, and also stayed on his game farm in Beaufort West, S.A. He died of a severe asthma attack in September 2001 while on holiday in Cyprus. He wrote two autobiographies: *One Life* in 1969 and *The Second Life* in 1993. No, I didn't catch sight of him but I certainly knew of his existence.

One area that I did not have a chance to investigate, as it was off-limits at the time, was District Six. This was a former inner-city residential area of Cape Town inhabited by a wide variety of people. These included former slaves, artisans, merchants and other immigrants, among them descendants of Malay people brought in by the Dutch East India Company during the time of their administration of the Cape Colony. It was home to almost a tenth of the city's population. However, now it was being cleared by order of the South African government.

During the early part of the Apartheid era, District Six remained a cosmopolitan area. Near the docks it contained a mix of coloured (mixed ethnic race) black Xhosa and a smaller number of Afrikaans whites and Indians. In 1966 the S.A. Government declared it a white-only area, forcibly moved the inhabitants to a newly-built Cape Flats Township, 25 miles away on flat, barren land, and bulldozed all the buildings apart from two churches and two mosques. By 1982 more than 60,000 people had been relocated. The Government gave four primary reasons for their action: According to the Apartheid philosophy inter-racial interaction bred conflict. It was a slum, not fit for habitation. It was crime-ridden and dangerous. It was a vice den with gambling, drinking and prostitution. These were the official reasons but most people believed that the government wanted the land because of its proximity to the harbour, Table Mountain and the city centre. However, international and local pressure made re-development difficult and they were able to only build the Cape Province University of Technology and some police housing units. Since the fall off Apartheid in 1994, the government has recognised the claims of former residents and pledged to support the rebuilding of some of their houses and a few have moved in. A Museum of District Six was completed in that year.

Interestingly, before he left South Africa in the 1960s and

INTO AFRICA – WITH A SMILE

achieved International fame, jazz pianist Abdullah Ibraham lived near District Six and was a frequent visitor. He said it was "Where you felt the fist of Apartheid. It was the valve to release some of that pressure. In the late 50s and 60s, when the regime clamped down it was still a place where people could mix freely. It attracted musicians, writers, politicians at the forefront of the struggle…. We played and everyone would be there."

Finally the time for me to leave Africa came and, together with Devon and Bruce, I rode my bike to the docks, drained the tank, disconnected the battery and watched, heart in mouth, as the dock workers strapped my precious BMW to a pallet and then swung it up and over the side of the big ship and down into the hold. That was it. Nothing left for me to do other than fondly farewell Erik and Hannie and take my few meagre possessions aboard. As the ship drew away from the quay and I watched the magnificent outline of Table Mountain fading into the distance I had to wipe away a few tears. "Farewell Africa, you have given me an amazing experience that I shall never forget and taught me many things. Thank you".

With or without the help of Dr Barnard, Africa had certainly pulled at my heart-strings. Would I ever return?

On board the Blucher

The Blucher

Blucher from opposite bank

Lee Marvin and Roger Moore during a break on set of 'Shout at the Devil'

Another boat from the film 'Shout at the Devil'

Along the coast

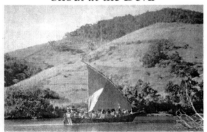

Dhow used in film

INTO AFRICA – WITH A SMILE

Diedre in East London

South African beach sign

Lil Collett
- South Africa WIMA Captain

In South Africa with largerphone

By a dam in South Africa

On the road

Riding an Ostrich

Table Mountain with 'tablecloth'

Cape Town harbour view

Cape Town University

Cape riders

In Stellenbosch

Stellenbosch house

Cape town riders

Hannie and Dierdre

Dierdre and Hannie

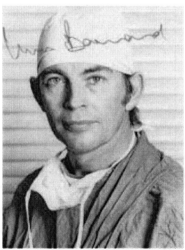
Dr Christian Barnard

Devon's bike being loaded

WIMA South Africa

Poverty is not an accident. Like slavery and apartheid, it is man-made and can be removed by the actions of human beings.
- Nelson Mandela

Loading bike onto ship with the boys

Linda's bike being craned on

Bike in the air

Goodbye to friends on the docks

Under way

Bruce and Linda on board

Bruce and Devon on board

EPILOGUE

Research for this book has given me a great deal of pleasure and information. I have learnt more about what was happening in Africa in the 1970s than I ever realised when I was actually there. When I was making the journey my priority was, understandably, to keep the bike going and myself upright on very bad roads. I was politically unaware and had no idea of what was going on around me in this respect: the social and economic organisation and who were the leaders of the countries through which I travelled. I have, therefore, tried to include these aspects and give an overall picture to make this not just a motorcycle travel story but place it in the context of its time.

To see the photographs from this book in colour format please go to Linda's website. www.lindab.id.au

PREPARATION

In those days we did not have the benefit of the internet to give information (true or false) on current or historical affairs or even the state of the roads. There were no books on Africa like *Lonely Planet* or *Rough Guide* and the RAC and AA were themselves not well-informed. Michelin maps were the best source of information but could not be kept up-to-date. Routes constantly changed with political blockages, new roads being built or weather conditions. One could say that I was blissfully unaware of what I may encounter and to this day I prefer things this way. Too much information may lead to preconceived ideas and worries. Things usually work out. For example, before I left I had the prescribed inoculations and took a first-aid pack but they did not prevent me from contracting malaria or dysentery. I survived anyway. I don't recall having travel insurance and of course there were no mobile phones or GPS systems.

I also think that I now have more opportunity and ability to mix with indigenous people because, since that journey, I have learnt to play the tin whistle and take it wherever I go. Although I had my lagerphone and singing voice, I did not use them apart from when I was with my own kind. Now when I am stuck at a border crossing or in other circumstances when I am kept waiting I pull out my instrument and play a tune. It is like a human magnet, for people of all nations relate to music and instantly smile and move toward me. It is a wonderful tool to meet people and I wish I had known how to play it then. Also, although I am a naturally friendly and gregarious person, I did not have the benefit of my later studies in Anthropology and consequent better understanding of other cultures which I can now apply to my travels.

It is always easy to have hindsight and wish that we had done things differently. I think I made the best of this journey. It would have been great to spend more time in each country but we are always limited in these overland epics by weather patterns, visa lengths and, of course, our finances. The journey depends on these

Into Africa – With a Smile

factors and our overall aim and even that can change on route.

The Motorcycle and Equipment

In this day and age, motorcyclists who intend to undertake an overland journey such as this have a great deal of makes and models to choose from, many specifically built for off-road terrain and there is a huge market in travel equipment. In the 1970s there was no such range of choice. A motorbike was a motorbike. If you wanted to use it for any other purpose than just normal road travel you changed the tyres and gearing and maybe the handlebars and suspension.

British bikes such as Triumphs and BSAs etc were all used in various areas of competition as were the 'Johnny- come- lately' Japanese machines. BMWs were renowned as touring bikes. Their shaft drive and horizontally-opposed engine made them very easy to maintain and, unlike British bikes, they didn't leak oil! They were used as police and missionary bikes in many countries and their German engineering standard was universally admired.

My bike, a 1957 R50 was already 17 years old when I bought it and I don't recall how many miles it had done or even how much I paid for it (probably about £200) but to me, already used to BMWs, it was perfect for this use. It had a very low seat height and centre of gravity so was good for balance. It had no electric starter, which kept the weight down, and meant that it didn't rely on the battery to start and only used the 6v unit for lights. They weren't too bright but I hardly ever travelled at night. The wheels were both 18" so I could use the spare inner-tube for either. It was quick to take out the bike wheel with the aid of the built-in stand on the back mudguard and the rims were easy to work with when changing a tyre. It used approximately 3 litres of 30 grade oil throughout the whole engine, gearbox and final drive and needed only a limited amount of spanner sizes to work on engine parts.

The bike did approximately 50mpg and the especially-made tank held 6 gallons. It could do 100mph if really pushed but was very happy at 70mph. This speed was not often used in Africa and it could tootle along at low speed in top gear. The brakes were drum and never very effective but, again, this was not too much of a problem at my rate of travel. When well-tuned, it was easy to start with the kick start as the engine was only 7 to 1 compression and could therefore run on low-grade fuel.

I had leather panniers bolted on. They weren't lockable or very big but adequate. The bike being so well-balanced, I could put quite a load on the back rack: a top box and my tent bag. The Harro tank bag had a separate leather compartment on the base where I kept my tools and spare cables etc and a zipped canvas compartment above in which my sleeping bag fitted.

My tent was a Marachal Pedestre, a ridge tent which was so well-made I still use it, on occasions, today (40 years later!). I used an Optimus petrol stove with a small pan and these were stored in the top-box with other kitchen utensils and any food I may carry. My sleeping bag was a Blacks down one with a cotton inner-sheet and I used an inflatable Lilo mattress. It was before the invention of Thermorest self-inflating, much lighter, mattresses.

I left UK with my two-piece touring leather and a Barbour jacket and over-trousers. I took a small dress, 2 T-shirts, 3 pairs of knickers, a towel and some toiletries. The leathers and over-trousers were stolen in Bangui together with my dress, a T-shirt and some underwear. I had a small first aid pack with anti-biotics, tummy pills, plasters etc.

INOCULATIONS

I went to the London Hospital of Tropical diseases to have smallpox, typhoid and hepatitis injections before I left and these were noted on a medical certificate which had to be produced at borders. I did initially take anti- malaria tablets.

CARNET DE PASSAGE

As described, I obtained one of these very important documents from the AA before I left and it was duly stamped in and out of every country and, when the bike was brought back to UK and the last paper signed, I received my bond back.

FINANCE

These were the days before the advent of the Euro or even bank cards. Every country I went through needed its relevant currency and, in many cases, it was illegal to take that money out of the country so you had to make sure you had spent or changed it before you left. We had to make a currency declaration both entering and leaving many of the countries. It was quite normal (though of course, illegal) to change money on the black market and it was quite easy to find this source in most cases.

I think I took about £400 with me, some in travellers' cheques (which proved hard to change) some in Sterling and some in American dollars. At that time the American dollars were the most popular form of currency.

WRITING THE BOOK

The material and inspiration for writing this book came when I found my notes about the journey that had been stored, in various countries, for forty years. The notes were by no means comprehensive, as my original diaries had not survived, but there was enough information to make me want to investigate further and also see if I could find and contact some of the people who were involved in the story.

For background on the political and economic state of the countries at that time I used the Internet and Wikipedia was a great help. The events that I personally witnessed made a lot more sense to me when I understood the overall picture. Many of these countries that I travelled through had only become independent from their colonial masters during the last 10-20 years. (Some, such as Rhodesia and South Africa were still under white rule.) It was understandable that they were finding their feet and, as in the extreme case of Uganda, in considerable turmoil. As I have said, it was probably a good thing that I did not know what could befall me.

I have tried very hard to contact other people who played a part in this story and, in some cases have been successful and they have been very generous in helping me sort fact from fiction in my own accounts. The journey took place 40 years ago and, as we all have differing memories of the same events, it has been very entertaining

LINDA BOOTHERSTONE

trying to reconcile our recollections and I am immensely grateful for their efforts. It has been like fitting the pieces in a giant jigsaw puzzle, difficult but a great deal of fun.

On a personal level it has been very rewarding to reconnect with people I hadn't seen or spoken to for decades.

My sincere thanks for their memories go to:

<u>Jacky Griffin</u>: She was my good friend who travelled on the blue truck. After the bus trip, Jacky did more touring in Rhodesia and South Africa and then returned to UK where she made a career in social work. She is now retired and living in Hampshire. (for another story which includes Jacky read *Three Wandering Poms*)

<u>Gene Visser</u>. He helped me re-build my bike in Durban. Gene married and had a farm in South Africa before moving with his wife and child to New Zealand where they now live.

<u>Devon Dold</u>. I met him in Cape Town before travelling to the UK on the Pendennis Castle. Devon and his friend Bruce stayed for a while in the UK before returning to South Africa. Devon later moved to Australia where he now lives with his wife. They have three sons.

<u>John Morgan</u> travelled through Africa in a Combi-van with Mike O'Connor. John ended up in Australia with Diana (Shilly) Shillabeer after the journey. He and Diana married, spent a while in the UK and then returned to live in Australia where Diana continued her career as a nurse and John became an investment banker. They now live in South Australia and have four, now adult, children. Mike stayed in South Africa working on projects as an electrical engineer and continues to live there.

<u>Jane Matthews</u> worked as a nurse with Diana Shillabeer in East Africa and I met them both in Kenya. Jane returned to Australia and continued nursing. She is now Jane Thomas, and lives in Victoria. She has five children and two grandchildren.

Thanks also go to the following for their unwitting involvement!:

At this point in time I have been unable to contact any of the other people whom I would like to thank ie Terry Wilkinson the leader of the blue truck. If anyone who reads this book recognises themselves

INTO AFRICA – WITH A SMILE

or any others mentioned please contact me.

I saw Kenneth Tilley very briefly in the UK in 1989. I understand that he was ill with a brain tumour and I have not been able to obtain news of him since.

Bruce Stephens now lives with his family in Zandvoort, Netherlands.

Les Fleming and his girlfriend, Eddy, with whom I lived in Durban, remained in South Africa where Les built up his electrical business. They had several children. I last saw them when they visited Spain in about 2002. I have been unable to contact them since.

Recommended reading, films and musicals

Books:

Uhuru	Robert Ruark 1962
	Buccaneer Books
Shout at the Devil	Wilbur Smith 1968
	McMillan
Facing Mt Kenya	Jomo Kenyatta
	1938 Secker and Warburg
Long Walk to Freedom	Nelson Mandela 1995
	Little Brown and Co
111 days in Stanleyville	David Reed 1965
	(Reprinted 1988 as *Save the Hostages*)
A Walk in the Night	Alex La Guma
	Mbari Publishers, Nigeria
Buckingham Palace District Six	
	Richard Rive 1987
	Ballantine Books
Out of Africa	Karen Blixen 1937
	Putnam (UK) Gyldenal (DK)
A State of Blood	Henry Kyemba 1977
	Ace Books
Hold My Hand I'm dying	John Gordon Davis 1967
	Chivers, Bath

189

LINDA BOOTHERSTONE

Musicals:

Ipi Tombi 1974 by Bertha Egnos Godfre and daughter Gail Laker. This starred Margaret Singara and toured South Africa, Nigeria, Europe, UK, USA and Canada. I saw it is London in 1976. Brilliant!!!

District Six the musical by David Krame and Taliep Petersen Premiered in SA in 1987 and shown to many audiences. Re-written by the author and re-toured in 2002

Films:

Gorillas in the Mist 1988. The story of Dian Fossey starring Sigourney Weaver. Nominated for 5 Academy Awards.

Out of Africa 1985 starring Robert Redford and Meryl Streep.

Shout at the Devil 1976 starring Lee Marvin and Roger Moore.

The Last King of Scotland 2006 starring James McAvoy and Forest Whitaker

Mandela: Long Walk to Freedom 2014 starring Idris Elba

Victory at Entebbe 1976 made for TV starring Anthony Hopkins, Burt Lancaster, Liz Taylor, Richard Dreyfus, Kirk Douglas

Raid on Entebbe 1977 made for TV starring Peter Finch, Charles Bronson, Yaphet Kotto

Blind Justice 1988 based on book Hold My Hand I'm Dying starring Christopher Cazenove, Oliver Reed, and Edith Brychta

Lion King 1994 Disney movie

WIMA Member of the Year 1975 – Linda Bootherstone
with other members in the UK.
Award presented by Gerry Clayton of Motor Cycling UK

Home again

LINDA BOOTHERSTONE

An African Odyssey

Finding facts to start the journey
Equip the bike and say farewells
Leaving the shores of mother England
What lies ahead no-one can tell

Morocco with its wool jelabas
Spice-filled markets, mountains high
Through the snow then down to desert
To palm trees and a star-filled sky

Sahara sand is next to battle
Corrugations do their worst
Blue-clad Tuaregs driving lorries
Help transport a bike that's bust

Leaving camels and silver crosses
Black figures now appear instead
Smiling ladies in coloured cotton
Balance baskets on their heads

In a land of changing conflicts
Fort Lamy is the place to be
Where fighting men from round the world
Give the French Foreign Legion their loyalty

Camped beside the Ubangi river
Prized possessions go astray
But heading on into the jungle
Butterflies flit and monkeys play

As rain falls down the mud gets deeper
Log bridges span the swollen streams
Each day is a constant battle
And bitumen roads are in my dreams

Into Africa – With a Smile

Are there gorillas in the mist?
In volcanic country with mountains high
Serengeti's wildlife banned for BMs
But now sealed road I can espy

At City Park the travellers gather
There's many a tale that's tall and true
With the jungle dangers now behind
It's southward to adventures new

Mt Kenya conquered, the east coast beckons
Coral reefs and graceful dhows
Pole Pole is a haven
But there malaria lays me low

Meershaum pipes in Tanzania
Then elephants and a copper mine
Lusaka police are causing trouble
But I escape without a fine

Between black and white the border runs
The smoke that thunders so they say
Uneasy peace on either side
While rainbows shine amidst the spray

At last Rhodesia smiles a welcome
Southern Africa a change indeed
A broken valve then halts the progress
But soon is fixed by friends in need

A job at last but not for long
With BM fixed I take the road
To find the city of gold and diamonds
And to its portals I am towed!

Durban dazzles with its friendships
Pilot boats and breakfast runs
BM rebuilt, I leave with sadness
To forsake the laughs and fun

LINDA BOOTHERSTONE

Can I now shout at the devil?
As my trip is near its end
Like an ostrich ignore the future
And put my head down in the sand

Table Mountain shouts the message
A ship awaits me at the quay
Now time to bid Africa adieu
Will I return another day?

THE AUTHOR

Linda Bootherstone has spent most of her life exploring the world by motorcycle. She is now based in Port Lincoln, South Australia where she follows her other interests of art and music but also takes trips on her motorcycle whenever possible.

She has written three books previously and made several recordings of her own songs.

To contact Linda, please go to her website (page 183) or email her at: casalinda2006@gmail.com

OTHER PUBLICATIONS
BY
LINDA BOOTHERSTONE

Daisies Don't Tell – An Illustrated Anthology of Poems
Linda Bick – Xlibris – 2010
ISBN: 978-1-4600-2352-8

Where Angels Fear to Tread – Travel Autobiography
Linda Bootherstone – Second edition 2015
ISBN-13: 978-1511561822

Three Wandering Poms – Travel Autobiography
Linda Bootherstone, Jacqueline Griffin,
Angela Griffin – 2014
ISBN: 978-1-5003-6716-9

CPSIA information can be obtained
at www.ICGtesting.com
Printed in the USA
LVOW10s1326280717
543001LV00018B/429/P